Sustainable Water Ecosystems Management in Europe

EU Report Series

Sustainable Water Ecosystems Management in Europe

Bridging the Knowledge of Citizens, Scientists and Policy Makers

Carlo Sessa

Publishing
London • New York

Published by **IWA Publishing**
Alliance House
12 Caxton Street
London SW1H 0QS, UK
Telephone: +44 (0)20 7654 5500
Fax: +44 (0)20 7654 5555
Email: publications@iwap.co.uk
Web: www.iwapublishing.com

First published 2012

British Library Cataloguing in Publication Data
A CIP catalogue record for this book is available from the British Library

Library of Congress Cataloging-in-Publication Data
A catalog record for this book is available from the Library of Congress

ISBN 9781780401140 (Paperback)
ISBN 9781780401157 (eBook)

Contents

Foreword

While science provides relevant and robust input for policy making, the communication of science to policy makers and the acceptance of the wider public of policy choices, made on the basis of scientific evidence, becomes a primary task for those interested in promoting public engagement in the research/policy cascade. Neither the popularization of science nor science education approaches are appropriate, because the objective is not to educate the public, but rather to engage it with the research products, as well as to make it part of the policy formulation process. New forms of dialogue and new mechanisms for introducing scientific issues to an audience spanning from managers and policy makers to the lay citizens, call for communication processes that enhance the link between the scientific processes, the policy products and the audience.

The European Union has embraced the concept of public participation and stakeholders' involvement, and all recent European directives explicitly require that Member States engage all stakeholders in all stages of the implementation. In the era of digital revolution permeating all aspects of life, great challenges and opportunities arise for the communication of science in contexts of public engagement, which allow building trust and broadening support while at the same time generating ideas and question paradigms. The Commission routinely runs public consultations on dedicated websites (see, for example, in the field of environment: http://ec.europa.eu/environment/consultations_en.htm). However, further steps need to be taken to improve the process of knowledge brokerage at all levels.

The EC 7th Research Framework Program funded project AWARE has taken one step forward, engaging a panel of randomly selected citizens living in three different coastal areas of Europe – the Gulf of Riga in the Baltic Sea, the southern North Sea coast and three adjacent river basins (the Seine, Somme and Scheldt rivers), and the Sacca di Goro lagoon in the Po river delta – in a pilot experience of knowledge brokerage with water scientists and decision makers focused on coastal waters quality. The knowledge brokerage process has been implemented over a full year, with workshops and public conferences at European and local level, and has been monitored and evaluated by a team of social scientists, to identify strength and weaknesses of the experience. Lessons have been drawn for its replication and transferability to different coastal areas and/or sustainable water management tasks, and beyond, to other sustainability research and policy issues.

A full account of the AWARE project experience and its transferability is provided in this book, together with the final declaration delivered by the citizens themselves at the last European conference in Brussels on 9 June 2011. The declaration is presented in the Appendix, but we suggest that reading of this book starts from that, as it gives the immediate impression of what lay citizens engaged in a structured knowledge brokerage process related to a complex sustainability issue may deliver. Besides some citizens' recommendations which address specifically the water management context, their more general recommendations to enhance the connectivity between science, citizens and policy making are worth of careful consideration. Particularly important is their quest of "scientific ambassadors", a new kind of experts endowed with the ability to interpret and transfer scientific evidence to the citizens and to public officials in charge of deciding about complex sustainability topics.

AWARE has shown that to enable citizens engagement in complex sustainability issues is not necessary that lay citizens acquire the knowledge and skills needed to carry on the needed investigation (this is up to the experts); what is required instead is that they have the ability to judge of the bearing of the knowledge supplied by others upon common concerns, providing that the scientific knowledge is transferred to them with adequate approaches as those experienced in AWARE.

We suggest therefore, in the last chapter of the book, ways to develop and expand the application of the AWARE approach, in the water sector and beyond, to other sustainability challenges Europe is facing, thus continuously creating communities of purpose to address common sustainability concerns. In its deepest and richest sense, a community must always remain a matter of face-to-face intercourse, and this is in a nutshell what working with the panel of 30 citizens in the AWARE project has demonstrated. In this respect, visual connectivity trough Internet technologies is not enough: citizens need to be engaged in mission-led processes, and feel that they are in charge and can objectively affect policy formulation and decision making, rather than simply bearing their consequences.

"Mind the gap" between global and local knowledge communities, and between the citizens, the experts and the public officials, is ultimately the key message of the AWARE experience. The citizens conference method is there to form transnational groups of citizens and to enable interdisciplinary and trans-boundary processes of knowledge brokerage, where global and local, scientific and everyday knowledge are put together to devise solutions and create communities of purpose to cope with sustainability challenges across Europe. Organizing systematically European citizens' conferences might be the new medium to cross-over local and global perspective and knowledge, bridging different people from different places into a face-to-face interaction environment, a sort of local community simulation. This may help all of us to discover – as the philosopher John Dewey highlighted already more than eighty years ago – the "lap of immense intelligence" in which we have the privilege to lie.

Carlo Sessa
Francesca Somma

Chapter 1
Background and objectives of the book

Carlo Sessa
AWARE Project Coordinator, Institute of Studies for the Integration of Systems, Italy

1.1 INTRODUCTION: THE AWARE PROJECT

The project AWARE, funded by the European Commission under the 7th Research Framework Program, has developed and tested a new way of connecting scientists, citizens and policy makers to enhance the effectiveness of their participation and cooperation in the water science and policy assessment process. Connectivity is ensured by applying a knowledge brokerage concept at the EU and local level. A pilot knowledge brokerage test has been successfully realized, focusing on three coastal areas of Europe (respectively the Southern North Sea Coast and the river basins of Seine, Somme and Scheldt, the Gulf of Riga in the Baltic Sea, and the Goro Lagoon in the Po Delta area, in the Adriatic Sea).

The project started in June 2009, and ended in December 2011, engaging a consortium of 13 research, NGOs and policy institutions from 7 EU countries and the EC Joint Research Centre (Ispra). The project has been concretely performed by recruiting a transnational panel of final water users: 30 citizens randomly selected from pools of candidatures gathered in the three coastal areas of Europe (10 citizens were selected for each area, from applications received through Internet). The citizens' panel has been engaged in a number of workshops with scientific experts, stakeholders and decision makers to assess the best scientific knowledge available, the local water management practices and the EU water policy framework, and to formulate their recommendations for interfacing experts, citizens and policy makers, to achieve a better management of coastal environments in Europe. The citizens' conclusions, taking the form of *AWARE Citizens' Declaration*, have been presented and discussed with intervening experts, stakeholders and decision makers in the AWARE European Conference "Linking research to policy in the water sector", held in Brussels, at the European Economic and Social Committee (AWARE, 2011a).

Results and lessons learned from the project are presented in several deliverables, and summarized in the AWARE brochure, available at www.aware-eu.net. The project applied a citizens' conference methodology to assess social acceptance of EU research goals and findings, which was first designed and fine tuned in two antecedent EC funded projects related to sustainable cities (RAISE: www.raise-eu.org) and sustainable urban transport (MOVE TOGETHER: www.move-together.net). Only in the AWARE project, however, the method has been fully tested at European and local level, and subject to a fully-fledged evaluation activity.

The AWARE citizens' conference methodology engages panels of randomly selected citizens – that is *'European citizens' juries* that in principle may originate from all countries of Europe and all walks of life, but were restricted to the three case studies areas for the purposes of the AWARE project. Citizens' panels are asked to make a critical evaluation of research goals, outcomes and management options, focusing on their societal acceptance. The evaluation is undertaken by means of brokerage among different forms of knowledge – that is between the different scientific disciplines needed to understand complex issues, the citizens' everyday life ("non expert") knowledge, the different stakeholders' and decision makers experiential knowledge. One of the key characteristics of the AWARE method is indeed linking together first scientists and citizens, very early in the process, in order to provide a common understanding of the issue at stake and let the lay citizens gain confidence to discuss further the issues with other stakeholders, and form their own independent opinions based on a more complete information.

Bringing together scientists, citizens and decision-makers in knowledge brokerage activities – by means of structured citizens' conference processes where citizens' panels discuss and present their recommendations *("Citizens' declarations")* to policy makers and society at large – is therefore at the heart of the proposed approach. This is a new way of connecting citizens, experts, stakeholders and policy makers in order to:

- Share a common understanding and awareness of the complexity of environmental and societal challenges;
- Discuss how research and new innovative solutions may help to tackle those challenges – now and in the medium-to-long term future;
- Deliberate about how various research outcomes (scientific advice, new innovative solutions etc.) could or should be taken up by governments or citizens themselves, for example by incorporating sustainability into planning or adopting more environmentally-friendly behaviors, respectively.

The pilot experiences conducted so far, and ultimately in AWARE, show that the method works, delivering important benefits for all the actors involved:

- The scientists learn to communicate results in a tailored manner to citizens, as well as policy makers and stakeholders. They also can broaden their research interests by accepting new inputs, and discovering a public interest in their results, beyond the academic fora;
- The citizens quickly learn key environmental concepts, change their mind becoming aware of the complexity of the challenges ahead, and reflect on how to tackle them with more systemic approaches. They may better assess which policies would be needed for solving complex problems, and choose to support their politicians in tackling challenging decisions and policies, thanks to the better understanding and greater commitment gained throughout the process;
- The various stakeholders benefit from the insights and opinions of the scientists, citizens and policy community usually assessed in the more neutral context – that is not heavily influenced by special interests – of citizens juries;
- The policy-makers – through meeting the citizens and hearing their proposals – can share common visions on societal challenges that could not be solved with simple policies, would require systemic and long lasting actions to be implemented beyond electoral terms, and a deeper consensus and commitment of all actors involved. Policy makers also gain further confirmation that decisions successfully involving all actors affected, through increased awareness for instance, are more effective in their implementation and outcomes.

The specific issue addressed in the AWARE project was the deterioration of coastal waters in Europe, and how EU funded research and EU and local policies may help to reduce deterioration and achieve a good ecological status of coastal waters across Europe. This environmental goal is a core objective of the EU Water Framework Directive (European Commission, 2000).

This piece of EU legislation provides a coherent framework for the implementation of policies and the assessment of water quality across EU Member States, including river basins, transitional and coastal waters. The outcome expected from a consistent application of the EU water policy assessment process – illustrated in Figure 1.1 – is the achievement of good environmental (ecological and chemical) status and the related benefits at a proportionate cost, or the maintenance of the water body in a moderate quality status with reduced benefits, due to the disproportionate costs of achieving a better quality status.

In this respect, the AWARE pilot experiments can supply a model for future implementation of participatory assessment of the water policies, whereby lay citizens and stakeholders are involved together with experts and policy makers in AWARE-like processes. These may be especially helpful to assess the "proportionality" of costs against benefits in specific case studies circumstances, because the criteria to be used in the assessment are not uniquely technical, requiring instead the explicit consideration of social norms and values.

Indeed, by mainstreaming the AWARE approach in the water sector, a better multi-level governance of policy issues can be achieved linking central and decentralized levels across Europe, and ensuring a constant interface and brokerage of scientific and citizens'/experiential knowledge at all levels, to support the water quality planning and implementation processes. As highlighted also by the AWARE pilot experience, nowadays the *dialogue* between scientists, policy makers and the end users targeted by water policies (citizens and stakeholders) works – to say the least – unevenly across different countries and different levels of governance. One recommended solution is therefore to replicate the AWARE experience, engendering new informal process of consultation on key sustainability issues which can enable a more productive public-policy interaction.

In the water sector, such an awareness raising process could help to bridge the gap between the citizens' locally specific and experiential knowledge, perspective and understanding of the topics, that of the elected representatives, and the more technical perspectives hold by the water managers and the experts involved in the water policy formulation process.

However, the AWARE method can be applied also beyond the water sector, to address other societal challenges in different sectors (transport, energy, etc.). This would mean to recognize that there are indeed different ways to effect an interaction between scientists, policy (managers) and the public, while tackling sustainability challenges, as portrayed in Figure 1.2.

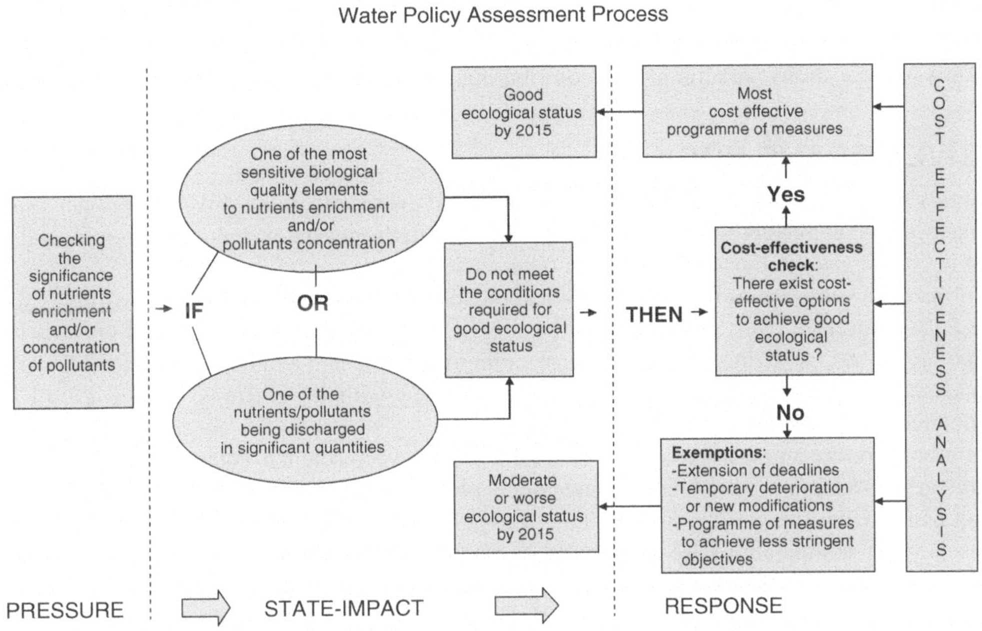

Figure 1.1 The EU Water Framework Directive assessment of coastal water quality.

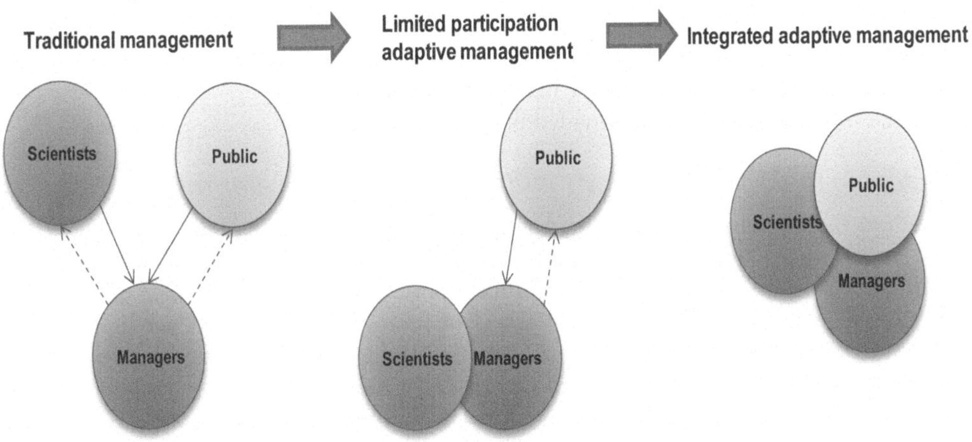

Figure 1.2 From traditional to integrated adaptive management.

The traditional way is to treat the three fields as entirely autonomous, interacting only within the established formal procedures of democratic societies, that is in the framework of public inquiries as prescribed by law or by delegation through elections.

Today it is however more common to follow what has been called a "participation-limited" adaptive management approach (Kusel *et al.* 1996) which supports the close interaction between scientists and policy but not with citizens or stakeholders. This is pretty much the approach that characterizes the EU Water Framework Directive, as the common framework built up to follow the implementation of the Directive is actively promoting a closer interaction between scientists and policy managers across EU Member States, through the setting up of several thematic working groups. But as the WFD strategists themselves admit, there are shortcomings in their approach, especially with respect to the use of research results from studies other than those commissioned by national governments; and with respect to inputs from stakeholders and civil society.

The integrated adaptive management approach takes a step further and tries to stage and learn from a closer interaction between science, policy, stakeholders and citizens. Such an approach is not necessarily always suitable, but it is increasingly needed and recommended to address complex sustainability issues. Dennison (2008) anticipated that the present major paradigm in coastal management is the shift towards sustainability. Tackling this paradigm successfully depends on the capacity of science to focus on environmental problem solving rather than problem analysis, filling gaps not only in the collection and analysis of scientific data, but also in common knowledge and information. AWARE has contributed to this paradigm shift through a key element of this process: improving science communication between scientists, key stakeholders and society. The method, as demonstrated in the pilot tests, displays indeed three main benefits: (a) it recognizes and uses the public as sources of information and knowledge; (b) it builds trust and broadens support; and (c) it generates ideas and questions paradigms thus contributing to both learning and creativity in problem-solving.

1.2 OBJECTIVES OF THE BOOK

The objectives of this book are to disseminate the methodology and the results of the AWARE project to a wider public of experts, practitioners, decision makers and lay citizens concerned with the coastal water quality issues, and beyond with the grand sustainability challenges the Europe is facing in other sectors.

The book scrutinizes the concept of bridging the Science-Citizens-Policy gap, showing how this is wider and deeper than the Science-Policy Interface (SPI) concept, as it engages early random selected panels of citizens in the evaluation of science and policy issues, with the aim to enhance communication and connectivity not only between scientists and decisions makers (managers and politicians), but also directly with the citizens, as we are all facing common sustainability problems and challenges.

The book proposes therefore a new approach to bridge the gap between science, society and policy making. It exploits the concept of integrated adaptive ecosystem management, presenting either theoretical foundations and emblematic practices, by engaging scientists, policy makers and the public – the latter including both stakeholders and lay citizens/water users – in comparable case studies of sustainable coastal waters management in different areas of Europe: the southern North Sea Coast, the Gulf of Riga in the Baltic Sea and the Sacca di Goro Lagoon in the Adriatic Sea.

The following chapters are extracted from the AWARE project deliverables and make a global synthesis of the project background, process, outcomes and evaluation. While presenting the background, additional attention is given to the concept of Science-Policy Interface and to theoretical foundations that may be found in political philosophy to the experience of engaging lay citizens at the interface of science with policy, that AWARE has piloted. The last chapter is devoted to the possible deepening of the approach to enhance public participation in water management, and its extension to tackle other grand societal challenges beyond the water sector. Finally, the book is completed providing in an Appendix the full text of the AWARE European Citizens Declaration.

A short abstract of the following chapters is provided below.

Chapter 2 – Developing the Science-Policy Interface to integrate water research and management

In chapter two, the authors initially discuss the recent change and steadily increasing interest in scientific contributions to policy making in the last 10-year period with reference to the IPCC and WFD and the increasing number of academic publications on the science-policy interface (SPI). Some frameworks and a couple of problem-areas for the science-policy linkages are also identified and discussed. The core of the chapter lies in 3 examples: firstly the experiences gained from the AWARE project in the three case studies on SPI, secondly, the recent development in science-policy linkages are shown for India taken from two recently completed projects. Finally, the recent initiative by the European Commission to develop a formal and operational SPI-group within the Common Implementation Strategy for the WFD is given. The chapter ends with a list of suggested actions to strengthen the science-policy interface.

Chapter 3 – Institutional and policy framework: environmental regulations applicable to coastal zones at EU and nationals levels

The deterioration of coastal waters, mainly due to the excessive inputs of nutrients from river basins, was the issue of concern in the AWARE project. Chapter 3 provides an overview of the policy framework regulating the quality of coastal waters in the EU including international regional sea conventions. It also illustrates the current status of implementation of the Water Framework Directive and the Integrated Coastal Zone Management recommendation in the Member States, as well as the institutional framework in the three AWARE case studies. Then, it analyses the state of compliance and the results of the present policy framework identifying the main gaps in the legislation. Finally, it includes some insights and perceptions on this topic from the AWARE experience.

Chapter 4 – Building a new Science-Citizens-Policy interface: theoretical foundations

The AWARE project adopted a pragmatic approach to develop and test a knowledge brokerage process engaging scientists, citizens and decision makers. This approach is mostly based on the citizens conference methodology, especially developed by the Danish Board of Technology and then applied more broadly. This pragmatic approach is extensively described in Chapter 5. In this Chapter, the focus is instead on theoretical foundations that can be found in two classical developments of political philosophy – the John Dewey analysis of the public and its problems and the John Rawls theory of social justice. These help to justify the extension of the Science-Policy Interface (SPI) concept to the broader notion of Science-Citizens-Policy Interface (SCPI), adopted as a new paradigm in the AWARE method. This analysis of theoretical foundations is needed to show how the methodology is robustly based on sound social research concepts.

Chapter 5 – The pilot experience with the AWARE citizens' panel

Following Chapter 4 focused on theoretical foundations of the AWARE methodology, this chapter provides a detailed description of the AWARE participatory experience. Indeed, a group of 30 lay citizens – the AWARE citizens' panel – participated in several workshops and conferences at the local and European levels, where they met scientists, policy and decision makers, and stakeholders concerned with the issue of coastal water ecosystem sustainability and methods of management. They were given a voice to elaborate a collective deliberation and sound recommendations on how to improve coastal water management in Europe. The chapter first describes the method and eventually relates the concrete process that led to articulate the local and the European scales, and to organize a real dialogue between a variety of actors.

Chapter 6 – The Future of the Gulf of Riga: pollution, water quality and fish production

In this chapter, the Gulf of Riga (GoR), one of the case studies chosen in the AWARE project is described in detail. The GoR waters are shared by five countries and thus being a transboundary water body, offers a major challenge to water resources management in the region. The regional differences in economy and political set up also add to the management challenges and the implementation of holistic mitigation measures in the GoR. These are some of the main reasons for lack of a good stakeholder engagement process involving different countries and also developing water management policies based on sound scientific inputs. In the AWARE project, attempts were made to look at the current gaps in the stakeholder involvement and the citizens-science-policy interface and how involvement and interfaces can be improved. This chapter provides a brief introduction of the geography, followed by a description of the eutrophication and fish stock fluctuations in the Gulf, which has been identified as one of the main problems in the region. The modelling approach used in the AWARE project will subsequently be described and finally the experience on the participatory citizens-science-policy process is discussed.

Chapter 7 – The diseased southern North Sea: current status and possible solutions

In the second case study of the AWARE project, namely the Southern North Sea and its three adjacent drainage networks (The Seine, Somme and Scheldt Rivers -3S-), the discussions have focussed on eutrophication of the coastal zone in relation to the excess nutrient delivered from land-based sources. Eutrophication at the coast is an economical problem especially in terms of fisheries and tourism. Despite the OSPAR (Oslo-Paris) Convention (1992) for the Protection of the Marine Environment of the North-East Atlantic and the European Water Frame Work directive (2000), the former aiming at reducing nitrogen and phosphorus by 50%, the latter dedicated for restoring a good ecological state of the continental water masses, eutrophication problems still exist. In order to feed the AWARE process linking citizens, stakeholders and scientists, and further explore possible measures for combating eutrophication, a chain of ecological models (a river model – SENEQUE/RIVERSTRAHLER- and a coastal zone model – MIRO) have been implemented using data available over a long period (1984–2007). Diffuse sources from agriculture and urban wastewater have been reconstructed, allowing quantifying the changes in nutrient deliveries resulting from changes in human activities in the watersheds and assessing their impact at the marine coastal seas. Whereas the OSPAR objectives have indeed been reached for phosphorus fluxes (effectively reducing the algal biomass, however above a threshold preventing damage effect of eutrophication such as foam accumulation of *Phaeocystis*), nitrogen flux did not change much. Several scenarios, discussed during the AWARE process, related to agricultural measures have been explored, and although they do not lead to a significant improvement of eutrophication at the coast, they have a significant effect for improving groundwater and hence surface water, a necessary condition for safely producing drinking water and restore the ecological quality of freshwater systems.

Chapter 8 – The Sacca di Goro: a cooperative decision making experiment for a sustainable lagoon exploitation

The Sacca di Goro is a relatively small and shallow lagoon located in the Southern part of the Po river delta. The lagoon is exploited for clam farming, which drives the whole local economy and support the social system. In the last two decades, many studies provided a robust scientific background on ecology, clam farming and environmental issues. A pilot decision support system was also implemented as a tool for supporting a sustainable lagoon exploitation. The lagoon is affected by pressures from both the Po river watershed and from local sources, also including shellfish farming itself. Macroalgal blooms and frequent anoxic outbreaks are among the most important threats, which have serious effects on clam production and damage the whole lagoon ecosystem. The main remedial measures are based on improving water circulation within the lagoon and water exchanges between lagoon and adjacent sea. Emergency measures also address macroalgal harvesting and management of clam faming during anoxic crises. The lagoon watershed is mainly exploited for agriculture, with possible impacts on the lagoon, which frequently cause conflicts with fish farming. The lagoon is also part of the Regional park of the Po river delta, which is mainly perceived as a constraint for the lagoon exploitation.

Within the AWARE project, a citizen panel was involved in a debate with scientists, stakeholders, end-users and policy and decision makers working in the lagoon and its watershed. The citizen panel then reassessed the scientific, policy and management information, in the light of their background, own needs and perceptions of policies and exploitation of the lagoon. The management of the lagoon and its watershed and the preservation of good water quality were thus identified as key issues for a sustainable use of natural resources and for a preservation of the aquatic ecosystem. In all the participatory events knowledge was provided by all the participants, in different formats and measure: expert knowledge mainly by scientists, tacit and local knowledge mainly by stakeholders and local policy makers. A pilot experiment consisting with the application of multicriteria methodologies facilitating the active participation of citizens, stakeholders, policy makers and scientists to the management of the Sacca di Goro has been also performed. The reaction of citizens displayed gradually within the process and led to a final declaration, that pointed out how the communication gap among actors is critical and a better knowledge of problems is fundamental for improving both citizen awareness or their participation in problem solving. Citizens also expressed the fear of wasting their time as they were suspicious of being involved in a process that may produce no results. This feeling poses a key point which is the effective willingness of all actors to participate in the decisional process, avoiding preventive lobbying. This can be seen as a sort of "cultural bottleneck" which tackles mainly scientists and policy makers.

Chapter 9 – Monitoring and evaluation of the AWARE pilot experience

This chapter describes the monitoring evaluation undertaken throughout the AWARE project, whose goal was to help consortium partners (coming from both scientific and social sciences) learn from each other and from interactions with citizens, stakeholders and policy-makers. The AWARE evaluation – better yet monitoring – accompanied the project throughout all participatory activities addressing both their design and implementation by looking at (1) the representativeness; (2) the structure; (3) the information used; and (4) the outcomes and deliberations arising from these participatory processes. Relevant information was collected by an evaluation team using observation, interviews, surveys, consultations with an advisory board engaged throughout the project, and discussions with the implementing partners. Chapter 9 discusses the lessons learned from monitoring the AWARE knowledge brokerage process and in conclusion provides eight recommendations for managers and policy-makers aiming to implement similar participatory processes to improve European policy development.

Chapter 10 – Enhancing connectivity between scientists and policy makers in the water sector and beyond

The goal of the AWARE project has been to demonstrate a new way of "connecting research, people and policy-makers in Europe to achieve sustainable water ecosystem management". The project used a variety of methods and activities to achieve this aim – including workshops, interactive conferences, online surveys and personal interviews. This chapter review the ex-post evaluation of the AWARE potential impact, as ascertained in particular in a number of keynote interviews and in a final ex-post evaluation workshop with water managers held in Brussels, at the end of the project. Future prospects to achieve a snowball effect, deepening the AWARE method application to foster public participation to integrated coastal zone and river basin management in the water sector and to expand its application to other sustainability challenges in other sectors (e.g. energy and climate, transport, etc.) are discussed. Practical lessons drawn from the AWARE pilot experience are given as a guidance for future participatory processes. The lessons learned are presented as tips about how better engage citizens, stakeholders and policy makers, scientists, and how to organise the knowledge brokerage process and deliver the outcomes in order to get effective results from the whole participatory process.

REFERENCES

AWARE (2011a). Final Publishable Report, EC FP7 Project, Brussels. Available online at http://www.aware-eu.net/

AWARE (2011b). Deliverable 4.3 – Bridging the Science Policy Gap: Best Practices in Citizens' Participation, EC FP7 Project, Brussels. Available online at http://www.aware-eu.net/ (accessed 10 April 2012).

Dennison W. C. (2008). Environmental problem solving in coastal ecosystems: a paradigm shift to sustainability. *Estuarine, Coastal and Shelf Science*, **77**(2), 185–196.

European Commission (2000). Directive 2000/60/EC of the European Parliament and of the Council of 23 October 2000 Establishing a Framework for Community Action in the Field of Water Policy, Brussels.

Jonathan Kusel *et al.* (1996). "Chapter 20: The Role of the Public in Adaptive Ecosystem Management", Sierra Nevada Ecosystem Project, David, University of California.

Chapter 2

Developing the science – policy interface to integrate water research and management

Per Stålnacke, Udaya Sekhar Nagothu and Philippe Quevauviller

2.1 INTRODUCTION

In 2009, Professor Johan Rockström was awarded the title "Swede of the Year" for his work on bridging science on climate change to policy and society. He was also in 2010 ranked as the 2nd most influential person on environmental issues in Sweden. This example might indicate that the role of science and research-based results for policy development is being increasingly recognized in recent years. The process within The Intergovernmental Panel on Climate Change (IPCC) established in 1988 is probably the best and most recent global example on how scientists all-over the world has contributed to science-based results to management and policy change. In the field of water resources management, the implementation of the Water Framework Directive (EC, 2000) is probably the best example of science based inputs to policy development within the European Union (EU). For example, many scientists throughout the EU have contributed to almost 30 guidance CIS-documents on for example, typology, classification, inter-calibration and river basin management plans (http://ec.europa.eu/environment/water/water-framework/facts_figures/guidance_docs_en.htm).

There are also an increasing number of academic publications on the science-policy interface (SPI) since the last 10–12 years (e.g., Spilsbury & Nasi, 2006; Gooch & Stålnacke, 2010; Quevauviller 2010). Several of these publications have stressed that a major problem in water management is the incorporation (or lack of) of scientific information into the policy and management processes (Gooch & Stålnacke, 2010) and that SPI seems to be the source of much frustration and misunderstanding (Roux *et al.* 2006). More specifically, scientists complain that policy makers and managers do not take into account the scientific knowledge generated, whereas, the policy-makers on the other hand blame that the scientists do not provide the kind of knowledge necessary to solve the problem of concern (Gooch & Stålnacke, 2010). Another dimension to this dilemma is that scientists tend to focus on more long-term problems that can be solved, or at least analysed, through the methods and methodologies that they are trained to use (Gooch & Stålnacke, 2010). Policy makers and managers, on the other hand, usually need practical answers to immediate problems which can shift over a very short time period. Policy makers, in their role as politicians or elected officials, also need to take into account the perceptions of their electorate, the people who hopefully will vote for them again the next time around (Gooch & Stålnacke, 2010). Thus, the differences between these two ways of formulating and understanding problems is significant and sometimes leads to claims and frustrations that the other side lacks a genuine interest in understanding the others perspectives. Despite this fundamental problem, the important role of science in policy-making has gradually become more accepted (e.g., the climate change scenarios). Scientists are also more aware of their role and at the same time funding agencies like EC require a strong policy-link in each project. More specifically, within the EU, funding is being provided to the scientific community to strengthen and develop methodologies for upscaling the Science-Policy Interface (SPI) and also the impact after the project has been emphasized. The dilemma is the fact that the career of scientists largely depends on their scientific output – which normally is counted as the number of publications in peer-reviewed academic journals preferably with high impact factor – and only to a much less extent on their contribution to policy measures (Stålnacke *et al.* 2011). From the policy and management angle, policy-makers and decision takers are gradually recognizing the need for a better interaction with the scientific community to the extent that platforms are being developed for frequent interaction between scientists and policy makers.

Another difficulty is to consider mandatory policy obligations on the one side, and the technical feasibility and scientific knowledge on the other, and reflect whether and how these can be properly interfaced (Quevauviller, 2010). The SPI in a water resources management context is, however, one of the most complex challenges that scientific and policy-making communities are facing. It involves knowledge sharing and exchange of information among a wide range of disciplines and actors as well as multiple governance system in a drainage basin. In many instances, the lack of proper communication, and of a coordination mechanism leads to research outputs not being used or simply known by policy-makers, and to policy research needs not being communicated timely to the scientific community. This issue has been widely discussed over the past few years in a number of studies (see e.g. Quevauviller *et al.* 2005; Quevauviller, 2007; Quevauviller, 2010; Gooch & Stålnacke, 2010; Stålnacke *et al.* 2011). It is only in recent years, that several initiatives have been taken up within the EU to reduce the gap between science, policy development and society. Thus, it seems it is still a long way to go before a well-established mechanism is in place and the gap between science and policy is reduced through civil society involvement.

In relation to this, it has also been pointed out a number of times that one of the major challenges facing the science-policy-participation interface in water management is the production, management and communication of data (Gooch & Stålnacke, 2010). It should also be noted that the development of environmental policies is a complex process, mixing legal requirements with issues of technical feasibility, scientific knowledge and socio-economic aspects, and requiring intensive multi-stakeholder consultations. In this context, the consideration of scientific progress represents one of the key aspects for the design of new policies and the review of existing ones. Currently, the SPI, in the water as in other sectors, is still unsatisfactory, to say the least. In fact, scientists view the policy end-user in the research project as the "legitimate" client for their research results, but on the ground there is a significant lack of transfer mechanisms that would allow passing the relevant information on the stakeholder and policy maker communities. The latter is often bounded by administrative, financial and/or time constrains in its capacity to translate research results into policy; often timely access to specific technical journals and thus relevant information is not readily available, so that relevant findings do not exit the tight circle of the specialised scientific community. There is also communication difficulties linked to the "jargon" used in the different communities. One of the problems identified is that research and policy have different timing concepts: policy tends to operate under the umbrella of a reactive philosophy, often under the pressure of short-term administrative mandates and long-term lobbying; science, on the other side, generally develops with an eye on the long-term perspective.

In addition, in most instances policy tries to achieve an acceptable compromise, whereas the scientific community strives for obtaining objective, scientific facts and wants to understand a phenomenon to the greatest possible detail. Last, but not least, career-reviewing schemes in science rarely give credit to effort for the integration of knowledge in fulfilment of policy objectives. As a result, scientists are not sufficiently motivated to perform such type of work, and in most cases, they may be not aware about the specific research issues that could, at a certain time, effectively support policy development. However, it is now well recognized that an improved understanding of environmental problems requires deeper awareness of multi-disciplinary scientific developments. Awareness by itself is, however, not sufficient; a stronger integration of research knowledge and results are also required at the various stages of policy developments (design, development, implementation and review). In many instances this is still far from being the case, due to a lack of a clear coordination mechanism between RTD and policy making.

In this chapter, we claim that there is a need for a stronger SPI framework. A sound scientific evidence base is required for the development and implementation of environment policies. In the water sector, international water policies rely on exchanges among policy-makers, stakeholders and the scientific community. This is well illustrated by the way European water policies, in particular the Water Framework Directive (WFD), are evolving, with on-going efforts to better organise SPI (Quevauviller, 2010). The need to develop a conceptual framework for a SPI related to water is at the heart of the discussions (see e.g. Quevauviller *et al.* 2005; Quevauviller, 2007; Gooch & Stålnacke, 2010; Quevauviller, 2010; Stålnacke *et al.* 2011). In this respect, lessons learnt so far show that interfacing science and policy is much more complex to achieve due the complex societal interests in water resources.

Based on this, we present various examples of SPI taken from the AWARE-project (www.aware.eu.net), a pan-European example from the WFD SPI initiative and from two projects in India. The chapter ends with a discussion about generic issues from the cases and suggestions for the way forward for the development of a science-policy interface to integrate water research and management.

2.2 THE SCIENCE-POLICY INTERFACE IN THE 3 AWARE CASES

A description of the Gulf of Riga, the North Sea, and Sacca di Goro cases dealt with in the AWARE project and its main issues or conflicts is available in Chapters 6, 7 and 8, respectively. An overview of the policy and institution framework in the three cases is found in Chapter 3. The three cases are at different stages of SPI development which could be justified due to their different political, economic and institutional set up that determines how governments prioritize SPI in their respective countries or regions.

2.2.1 The Gulf of Riga case

The Gulf of Riga is shared by two countries and hence adds to the challenges in addressing the SPI issue. A basic component is the presence of a transboundary water management authority that could take the main responsibility for developing a good SPI. Such an authority is not present in this case, that has implications for overall monitoring and management. In addition, the national governments sharing the Gulf of Riga have other priorities other than SPI to deal with, especially due to the recent economic crisis. The lack of a complete assessment of the water ecosystem of the Gulf of Riga was identified by the AWARE team as one of the main constraints in the science-policy linkage. In particular, the region lacks sufficient scientific data (especially on the impacts of water pollution on habitats) and irregular funding for monitoring, except for some projects funded by the international donor agencies and the EU. During the early 2000, there were several short-term projects funded by Sweden and Denmark, for example, to assist Latvia and Estonia to comply with the Water Framework Directive (WFD) both from a scientific point of view as well as capacity building for the water management authorities. These projects also worked to enhance coordination between the different ministries and agencies working in the water and environmental fields. It appears that until now the role of scientists has been rather passive concerning water management planning. Although they contribute knowledge, they are rarely involved in the decision-making process. Recent initiatives however, show that research projects are putting emphasis on sharing data and providing inputs to policy development, to address the science-policy gap. More precisely, during the development of the first River Basin Management Plans for the WFD, the Ministry of the Environment in both countries established working groups/and coordinating committees from different stakeholder groups for each river basin district. Coastal and transitional waters were assigned to specific river basin districts. The stakeholder engagement process started in 2005/2006. The groups/committees met regularly to discuss the progress of river basin management planning. Information on the draft river basin management plans was published on internet. Wider public consultations with stakeholders were launched in the end of 2008 and continued into the first half of 2009. The river basin management plans are available online and open to public (www.lvgmc.lv and www.envir.ee). Although public hearings were organised, it must be stressed that the participants of these water related meetings were representatives of stakeholder groups, and not certain to what extent they represented the citizens. It is apparent that a relatively "top-down" approach prevails in the two countries. However, local initiatives by NGOs are also common. Municipalities and NGOs are probably those institutions which are closest to the citizens and also those which are in the best positions for bringing up the voices of local inhabitants. Different stakeholder groups (NGOs, training centres, municipalities) also have taken actions to support the implementation of Water Framework Directive (WFD). For example, The Baltic Environment Forum (BEF) Latvia has been particularly active in increasing the role of NGOs in the implementation of the WFD and increasing public participation. The activities are project based, thus they are time and resource limited. Nevertheless, they have played important supplementary role in creating a better understanding on water management issues in the countries.

It is also interesting to note that within the AWARE project, the initiative for engaging the stakeholders in the Gulf of Riga case was assigned to the scientific community, which gave increased credibility to the activities, from the point of view of the stakeholders. Policy makers were interested in the process followed in AWARE due to the implementation requirements – to include both scientific information and public participation – specified in the Marine Strategy Framework Directive.

2.2.2 The North Sea case: Seine, Somme and Scheldt rivers

Whereas participatory mechanisms are nowadays currently observed between scientists and managers and decisions maker, for example Water Agencies or other public authorities through long terms interdisciplinary programmes, such as the PIREN-Seine programme since 1989, the involvement of scientists in the large public consultation campaigns is weak. The message from research, often very technical, is probably difficult to share with both the public and the executive organizations. It was during the AWARE-project in our consultation with stakeholders, suggested that scientific data should be popularized, for example, through media, direct marketing or even home mailing. It was also suggested to launch a periodical (like the EDF/Electrabel's magazines) in order to inform about water control and treatment (based on the results of delegate's report), to acquaint citizens with the current and coming actions and to offer them a right of reply. Moreover, the following was also discussed and proposed:

- Make it mandatory for the stakeholders (provider, distributor and operator) to publish once a year, providing clear information about the investments and expenses made for improvement of the water management situation.
- Consolidate and record the facts and figures of the water body in a centralized way, in order to identify any redundancy in data gathering.
- Provide access to each group to those elements in a way that they can understand: detailed and gross access for scientists, summarized form for stakeholders and decision makers and attractive rundown for citizens. The objective is to achieve the sharing of the same information by any involved actor, to allow citizens

to understand what is going on around them and to make them realize that "what I am doing here does have an impact there."

- Increase young generation awareness through school. It will have a domino effect, and it will permit to bring parents to think about the impact of their choices as consumer activists.

2.2.3 The Sacca di Goro case

The Sacca di Goro and the Province of Ferrara display a long tradition in adaptive ecological management through the active sharing of clam farmers' knowledge. Though, historically, the inputs of other stakeholders and citizens has not equally pursued. Fishermen above all have been strictly working with policy subjects since their interest on coastal water management is very much linked to their economic interests. The Province of Ferrara has a central role in the research and maintenance activities of the Sacca di Goro; it collects oceanographic data through an extensive monitoring network, coordinates the dredging activities with the support of the fishermen consortia and elaborates thematic GIS maps and other informational tools for the management of the Sacca di Goro.

However, ecological and environmental research is not sufficiently integrated with economic and social sciences to give definitive answers. EU, national and regional research projects are directed to improve our knowledge of the system but with continuous reduction of resources. A better comprehension of the system would avoid unnecessary conflicts and help to define common objectives. Moreover, the policy objectives are not clearly defined and social aspects are often not considered or they are managed without considering the "sustainability" (the lagoon as a "welfare" tool). Another hindrance is that the market is subjected to wide fluctuations and can drive resource exploitations as both high and low prices can result in high exploitation levels in addition to the need of high investments for maintaining or improve the actual production.

2.2.4 A cross-comparison of the 3 AWARE cases

Within the AWARE project one weakness was the difficulty to involve policy makers in the case studies. And even in cases where it was possible to organize some form of participation of policy makers, the nature of the interaction was not very productive – it remained on the level of political statements rather than producing a truly exchange of ideas among the policy makers and scientists as well as with the citizens. Unfortunately this appears to be a systemic problem rather than an organizational weakness of this specific project, as highlighted by the evaluation process in AWARE given in Chapter 9 (this volume).

Our analysis of the science-policy interactions in the three cases outside the AWARE project itself shows that the science-policy linkages primarily have been driven from individual research projects. This was particularly seen in the Gulf of Riga and North Sea cases. In Gulf of Riga it is apparent that NGOs play a crucial role in mediating scientific results and be a link between science-policy and citizens in terms of information dissemination. The clearest science-policy linkages could be observed in the Sacca di Goro case where one public institution were heavily involved in natural science data collection, and coordination activities for certain economic sectors (clam farming). However, the exact role and nature of the effectiveness of this science-policy platform was difficult to assess.

2.3 A CASE FROM INDIA SHOWING THE NEED FOR SPI

SPI concept is relatively new to countries like India, unlike in the EU where much emphasis has been given in the recent years to develop science based policies. As stated earlier, a good example in the EU is the Water Framework Directive that has been developed with scientific inputs from various disciplines. Policy development in India, like in other developing countries is by and large influenced by politics and economics of the relevant sector and not much of science. Stakeholders are merely informed about any new policy, but never engaged in its analysis and development. Such policies developed without stakeholder involvement have failed to address real problems. Stakeholder participation will not only enable stakeholders to identify their diverse objectives, flag problems and conflicts, but also contribute to their resolution (Slob *et al.* 2008). Hove (2006) emphasizes the necessary existence of 'negotiations' in participatory processes to effectively use plurality of approaches. Stakeholder involvement is also expected to yield two positive outcomes: realistic and more effective policies and plans, and improved implementation.

It is only since 2000 that emphasis on science based national programs and policies are gaining momentum in India. A number of initiatives have been taken recently that demonstrate the intention of the government to bring together scientists and policy makers to cooperate and work towards a stronger SPI, including the National Action Plan for Climate Change (NAPCC) and the National Water Mission of India launched in 2008 (http://moef.nic.in/downloads/public-information/presnt_CC.pdf). The NAPCC clearly states the need to organize research and that future climate adaptation strategy development should be based on well-grounded scientific recommendations. However, it neither specifies how this will be put into action nor provides guidance to strengthen SPI. In 2011, scientists from reputed institutes in India were nominated to be part of a national network to provide inputs to the forthcoming initiatives of the

Government of India. This is perhaps one of the first steps taken by the government to bring scientists closer to policy makers. More such measures are needed across the country, which has a federal political structure and diverse problems linked to natural resources management.

2.3.1 Climawater and Climarice projects

In this section, experiences showing the need for SPI from two projects in India, namely, Climarice (www.climarice.com) and Climawater (http://web.iitd.ac.in/~gosain/CLIMAWATER/index.html) will be shared. In the two ongoing projects in India (Climarice and Climawater), demonstrating how SPI can be strengthened at various administrative levels is one of the main objectives. The two projects have been attempting to test the climate change impacts on water and agriculture sectors, in the Godavari and Cauvery river basins in India respectively. The projects are in the process of identifying suitable adaptation measures together with stakeholders and farmers to address the climate change impacts. The projects so far have used several methods, including stakeholder workshops, Focus Group discussions, stakeholder panels, key stakeholder interviews as part of the stakeholder integration. These interactions were attempted at different stages of the project and at different administrative levels. The assumption was that stakeholder involvement can lead to a stronger SPI and a meaningful policy development and implementation. This was mainly done to expose the stakeholders to scientific outputs at an early stage and based on their feedback to revise the results or outputs. The following steps have been followed in the projects that demonstrate at a pilot scale how the SPI can be accomplished at various levels.

- Presenting the main scientific results to the selected stakeholder panel at the regional level (include senior water managers and representatives of Farmers organizations)
- Revising the results or outputs based on feedback from the stakeholders
- Developing policy guidelines based on the scientific results
- Presenting the policy guidelines to selected key policy makers at the provincial and national level
- Revising and drafting the policy recommendations that may be taken as inputs to the new policy document at the state and national level.

At one of the stakeholder panel meetings (held during August 2009) in Trichy district located in the Cauvery river basin in India, the regional Joint Director of Agriculture, Department of Agriculture, suggested that the District contingency plans developed each year should incorporate outputs from scientific projects such as Climarice. These contingency plans further provide inputs to the state level action plan, indicating that SPI can operate at different administrative levels. At the meetings, stakeholders in general were positive to SPI initiatives and insisted that scientists and policy makers should meet more frequently to exchange information especially when developing new policies.

One of the main outcomes of the Climawater project was a Stakeholder Forum that was established to promote SPI at the provincial level. The Forum will act as a platform where scientists and policy makers will meet in the future to discuss outcomes from ongoing scientific programs and channel the research outputs to relevant policies.

2.3.2 Lessons learnt from the India projects

To strengthen SPI, it is essential to involve stakeholders, including policy makers at an early stage. This will give some sort of ownership to them and also helps in building trust between different actors. Overall, policy makers are not interested in complex frameworks that they cannot implement. Scientists should understand the policy needs and accordingly conduct research to address the needs. Stakeholders' experiences are useful in improving the scientific analysis as we observed in our study. There is a growing interest within research community and stakeholders in India to co-operate on issues such as climate change and natural disaster management which are complex and demand inputs from scientific work.

In the future, more emphasis is needed to strengthen SPI through a well-developed institutional framework and resources at the provincial and national level. Policy makers should also recognize that research is time consuming and requires investments. The Government of India should provide special funding to strengthen SPI on the lines of EU 7th Framework program. Climawater and Climarice projects clearly demonstrate that SPI is essential especially in addressing future climate change impacts, in terms of assessment and effective adaptation at multiple scales.

2.4 THE SPI-APPROACH IN EUROPE (WFD)

The Water Framework Directive (WFD) (European Commission, 2000) provides a stable platform, which enables to build up best practice exchanges among different players (policy implementers, technology providers, scientific community, industrial stakeholders, NGOs etc.). As described above, it is recognised that efforts are needed to improve communication, that is presenting results of research and demonstration projects in an easily usable form by policy-makers. Adversely, the way research results are considered by the policy-making community is prone to improvements, in particular for tackling difficulties to integrate the latest research developments in legislation. These improvements were thought to be possible through the development of an operational "science-policy interface" based on a coordination of relevant programmes/projects with direct relevance to the WFD implementation (Quevauviller, 2010), including:

- ✓ An evaluation of the research needs linked to specific policy milestones (e.g. state-of-the-art scientific information or tailor-made research, R&D and demonstration);
- ✓ A mechanism to identify research projects which are most relevant with respect to policy steps, and which might be "validated" through demonstration activities with the aim to be disseminated and applied at the most appropriate level (regional, national or EU);
- ✓ An exchange platform enabling both scientists and policy makers to discuss the corresponding research and policy agendas from the very beginning in order to ensure a more structured communication at all appropriate levels of policy formulation, development, implementation and review.

This interfacing concept is shared by a growing number of actors involved in WFD issues, and discussions are on-going to identify concrete steps. The basic idea is that an operational mechanism should strive to establish strong links among different funding mechanisms existing at the EU and national level and thematic policies, in particular, new partnerships could be developed by combining various types of projects (e.g. research, demonstration, capacity-building, education, networking) funded by various programmes (e.g. RTD Framework Programme, LIFE+, COST, Structural and cohesion funds etc.) as well as projects funded at regional/national level, and promoting the establishment of innovative partnerships involving scientists, policy makers, managers and other stakeholders for the effective integration of science outputs into policy and management decisions.

The way research outputs have been integrated into WFD implementation has been discussed in recent publications (Quevauviller, 2010; Quevauviller *et al.* 2011). While research needs were generally well considered in successive EU Framework Programmes (FPs), projects could often not anticipate the way policy steps (e.g. monitoring programmes, programmes of measures) would be designed by Member States. This has sometimes resulted in scientific outputs which, although they had the potential, were not directly usable for policy implementation. In addition, the coordination among different actors has not been sufficient to allow a full integration of input from RTD projects into the policy discussions, neither did they make them accessible in time to meet policy needs. This was due to a lack of the above mentioned "science-policy interface".

Interactions among policy and scientific communities vary according to the phasing of research programming, project development and of course timing of policy milestones. To date, these interactions are not coordinated in a systematic way and are rather functioning on an *ad hoc* basis (as links among "individuals" rather than "institutional" structured links), which may create an obstacle in the pursuit of effective collaboration, and efforts are going on in Europe to improve this situation as described below.

2.4.1 A pilot initiative: the CIS-SPI activity

In response to the above considerations, the European Commission is developing a concept in the framework of an expert group co-chaired by the Directorate-General for Research and Innovation and the ONEMA (Office National pour l'Eau et les Milieux Aquatiques, France). This activity aims to establish an efficient and sustainable science-policy platform linked to the Common Implementation Strategy of the WFD and it includes the development of the so-called WISE-RTD web portal briefly outlined below.

The starting point of the expert group (known as CIS-SPI group, standing for CIS "Science-Policy Interfacing") is that the development of an operational interface would only be possible through interactions and guidance of a dedicated group mixing policy people, scientists and stakeholders. A pilot phase has been carried out in 2009–2011, providing insight on research needs, competences and gaps, notably on the basis of discussions with thematic CIS working group in their field of activities (e.g. ecological status, chemical monitoring, groundwater, floods, water scarcity and drought). The activity strives to develop a framework to improve exchanges between science and end-users for the transfer of research results and for the identification of research needs linked to the policy cycle, based on a few successful examples.

This pilot action, if successful, could help paving the way for a kind of "institutionalisation" of science-policy interfacing within the CIS, enabling to channel discussions, to improve efficient transfer of research information, to help Member States to get a timely access to scientific information supporting water policy implementation (in particular the Water Framework Directive and its daughter directives, as well as other relevant directives such as e.g. the Floods Directive) and to identify short, medium and long term research needs. To be effective, the entity should rely on national and regional relays, ensuring that scientific results and related recommendations are "translated" in the form of usable products at national and regional levels, with possible supporting demonstration activities. This will only be possible if Member States and regional implementers are willing to establish an "information chain" and put efforts into translation and research implementation. One objectives of such a SPI under the CIS umbrella is to ensure that scientific knowledge and technical solutions are efficiently transferred to policy advisors, regulators and managers, feeding the implementation timetable and further reviews of water policies.

Such an operational interface could dramatically improve the two-way communication between science/technical experts and policy advisors, regulators and managers and improve synergies between various water related tools and programmes. Supporting and strengthening the two-way communication between science and policy may have a tremendous impact. It may facilitate and speed up improved technical implementation of the WFD via problem-solving (through networking and regular knowledge exchanges) and creating solutions that might be too time-consuming or costly to develop in individual Member States, hence representing a significant economic impact. It may ensure that short, medium and long-term research needs are identified, classified according to different funding levels (EU, national or regional) and taken over by the scientific community (and other stakeholders, e.g. consultancies) so that the iterative river basin management planning process is constantly fed by relevant and tailor-made research outputs.

By gathering representatives from scientific, policy implementers and stakeholder communities (involved in relevant RTD projects, networks or working groups), a SPI group working at the CIS level would have to provide recommendations for the development of a framework in order to improve exchanges between science and end-users for the transfer of research results and for the identification of research needs linked to the policy cycle. The aim is to guide the cycle consisting of reviewing RTD projects with regard to their policy relevance, demonstrating their applicability, communicating their outputs, assuring dissemination and accessibility of their results, compiling feedback from users, and identifying further research needs linked to policy development, implementation and review.

Also, water policy is quite often at the crossroad of several issues, such as CAP reform, climate change adaptation and/or mitigation policies, biodiversity preservation or energy policies. Integrating these various aspects and science dealing with these issues is crucial for the relevance of future water regulations. A SPI group with the capacity to deliver a proper state of the art of existing scientific knowledge, potentially of interest to water policy implementers and stakeholders, and able to identify further research needs linked to key policy milestones on these aspects, would be of real added-value. This is the aim of the current initiative for which a mandate presently is being developed to the EU Water Directors.

A functional SPI at the European level will be even more successful if information and communication reaches out all the way to local water authorities and stakeholders, occasionally even below the level of competent authorities of the WFD river basins. In some countries "translator" institutes exist which may play a pivotal role. Additionally, much can be gained by coordinating (to some extent) research funding on various levels, in particular allowing quick response to emerging policy research needs. In particular at the European level, funding mechanisms which allow the initiation of research to answer policy needs (say within months) are currently lacking.

2.5 CONCLUSIONS AND WAY FORWARD

The AWARE experience, and in particular some of the interviews undertaken, show that there is a gap in the dialogue between scientists and policy-makers, as well as between policy-makers and the general public especially on complex topics that require a technical background. As for the first gap, the dialogue is often unidirectional, with policy-makers asking the scientists for advice but with scientists not always directing their research to answer policy questions. This dialogue works unevenly across different European countries as well as at different EU, national and local levels, and this fact calls for a better connectivity between the body of research produced across the whole European Research Area and the advice provided to EU, national and local policy-makers.

In summary, we propose the following actions to strengthen the science-policy interface:

- Early communication among scientists starting strategic research with policy implications and relevant advisors in the policy fields concerned. A better dissemination of scientific results that could potentially be used in the policy making and implementing process should take place (not only in scientific journals but also in a way that is 'readable' by policy makers) and adequate training on use of research outputs could be envisaged;
- Contacts and formalised discussion platforms for policy makers and scientists should be established in relevant fields to strengthen SPI. A sustainable dialogue should be established at regional, national and supranational (EC) levels, enabling exchanges between scientists and policy-makers which are not 'naturally occurring' for a meaningful SPI;
- Dedicated research in support of policies should involve policy-makers as well, that is networks have to be a mix of representatives from the scientific and policy communities;
- Finally, research aiming to support policies should develop a work plan mixing good science and demonstration of its applicability through testing against policy deliverables/milestones.

Moreover it is important that scientists have the capability and funding to take a broader view and to put their results into a broader political and policy context and identify what is good enough for policy as distinct from scientific publication. This will require a policy shift by the major funding organisations (national research councils, EC etc.) that presently almost solely "reward" academic publications in peer-review scientific journals. Policy makers on the other hand should be more receptive to science despite it might make "life too complicated'' and also invest enough time with the scientists.

REFERENCES

European Commission (2000). Directive 2000/60/EC of the European Parliament and of the Council of 23 October 2000 establishing a framework for Community action in the field of water policy. *Official Journal of the European Communities*, **L327**, 22.12.2000, 1.

Gooch G. D. and Stålnacke P. (eds). (2006). *Integrated Transboundary Water Management in Theory and Practice: Experiences from the New EU Eastern Border*. IWA Publishing, London.

Gooch G. D. and Stålnacke P. (eds). (2010). *Science, Policy and Stakeholders in Water Management: An Integrated Approach to River Basin Management*. Earthscan, London, p. 166.

Quevauviller Ph., Balabanis P., Fragakis C., Weydert M., Oliver M., Kaschl A., Arnold G., Kroll A., Galbiati L., Zaldivar J. M. and Bidoglio G. (2005). Science-policy integration needs in support of the implementation of the EU Water Framework Directive. *Environmental Science & Policy*, **8**, 203.

Quevauviller Ph. (2007). Water protection against pollution. Conceptual framework for a science-policy interface. *Environmental Science and Pollution Research*, **14**(5), 297–307, DOI: 10.1065/espr2007.06.432.

Quevauviller Ph. (ed.). (2010). *Water System Science and Policy Interfacing*. RSC Publishing, Cambridge, ISBN: 978-1-84755-861-9, p. 430.

Quevauviller Ph. (2010). Is IWRM achievable in practice? Attempts to break disciplinary and sectoral walls through a science-policy interfacing framework in the context of the EU Water Framework Directive. *Irrigation and Drainage Systems*, **24**(3–4), 177–189, DOI: 10.1007/s10795-010-9102–x.

Quevauviller Ph. (2011). Integrating science-based recommendations on water and climate change in river basin management planning – goal, opportunity or inaccessible dream? In: The Water Framework Directive – Action Programmes and Adaptation to Climate Change, Ph. Quevauviller, U. Borchers, C. K. Thompson and T. Simonart (eds.), The Royal Society of Chemistry, Cambridge, pp. 34–43.

Quevauviller Ph., Borchers U., Thompson C. K. and Simonart T. (eds.). (2011). *The Water Framework Directive – Action Programmes and Adaptation to Climate Change*. The Royal Society of Chemistry, Cambridge, p. 214, ISBN: 978-1-84973-053-2.

Roux D. J., Rogers K. H., Biggs H. C., Ashton P. J. and Sergeant A. (2006). Bridging the science–management divide: moving from unidirectional knowledge transfer to knowledge interfacing and sharing. *Ecology and Society*, **11**(1), 4. [online] URL: http://www.ecologyandsociety.org/vol11/iss1/art4/ (accessed 25 May 2012).

Slob A., Gerris L. and Ellen G. J. (2008). Sediment management and stakeholder involvement. In: Sediment Management at a River Basin Scale, P. Owens (ed.), Elsevier, Amsterdam.

Spilsbury M. J. and Nasi R. (2006). The interface of policy research and policy development process: challenges posed to the forestry community. *Forest Policy and Economics*, **8**(2), 193–205.

Stålnacke P., Gooch G. D., Nagothu U. S., Nesheim I., Barkved L. J., Grizzetti B., Rieu-Clarke A., Deelstra J., Thaulow H., Berge D., Lo Porto A., Nhung D. K., Manasi S. and Portugués S. B. (2011). Integrated water resources management: STRIVER efforts to assess the current status and future possibilities in four river basins. In: Transformative Research for Sustainable Development, C. Carlo, J. D. Tàbara and J. Jaeger (eds), Springer Verlag, Berlin Heidelberg, pp. 155–172, ISBN: 978-3-642-19201-2.

Van Den Hove S. (2006). Between consensus and compromise, acknowledging the negotiation dimension in participatory approaches. *Land Use Policy*, **23**(1), 10–17.

Chapter 3

Institutional and policy framework: environmental regulations applicable to coastal zones at EU and national levels

Camino Liquete[1], Francesca Somma[1] and Udaya Sekhar Nagothu[2]

[1]Joint Research Centre – European Commission, Institute for Environment and Sustainability, Ispra, Italy
[2]The Norwegian Institute for Agricultural and Environmental Research (Bioforsk), Norway

3.1 INTRODUCTION

Coastal areas provide crucial resources for wildlife (e.g. key nursery habitats), human well-being (e.g. recreation opportunities) and economy (e.g. fisheries). Some authors estimate that coastal ecosystems may contribute 77% of global ecosystem services value (Martínez *et al.* 2007). More than 200 million people, equivalent to 44% of the coastal Member States' population, live in the EU coastal regions (Eurostat, 2011). The increasing pressures and demand for coastal resources causes habitat loss and degradation, pollution, climate change and overexploitation of fish stocks, thus leading to the degradation of coastal ecosystems (EEA, 2010). The first reporting by Member States on the conservation of wildlife pursuing the EU Habitats Directive indicates that more than two thirds of coastal habitat types and more than half of coastal species have an "unfavourable" status. This degradation has large negative social and economic consequences.

The deterioration of coastal waters, mainly due to the excessive inputs of nutrients from river basins, is the issue of concern in the AWARE project. Nutrient enrichment in coastal zones leads to a number of undesirable effects, such as excessive algal growth, changes in the stability and structure of coastal ecosystems, toxicity effects in food from shellfish and beach closures affecting tourism.

Situated at the interface of land and ocean, coastal zones have long lain in a sort of legislative limbo, falling within the domains of legislation regulating either land or marine-based human activities. Thus, the concept of coastal zones as a seamless transition zone of the land-sea interface remains an elusive one, both in the scientific and policy contexts.

This chapter provides an overview of the policy and institutional framework regulating the quality of coastal waters in the EU, and the international protocols currently regulating the quality and the use of water in European regional seas. It also illustrates the current status of implementation of the Water Framework Directive and the Integrated Coastal Zone Management recommendation. This compilation was the basis for discussion at the AWARE stakeholder and citizens' workshops, centred on Europe's response to the challenges presented by all sorts of water quality deterioration in coastal areas.

3.2 THE CURRENT POLICY FRAMEWORK

3.2.1 Review of EU water quality policies relevant to coastal areas

In the last two decades, the European Union has adopted a number of directives which directly or indirectly aim at the protection of European waters through sound management. Where initially the approach was sectorial, later legislation has assumed a more holistic approach, including the whole ecosystem as a basis for regulation (commonly referred to as ecosystem-based management).

Sectorial approaches focused on specific issues (such as pollution abatement) by addressing emissions caused by single sectors (agriculture with the Nitrates Directive, industry with the Dangerous Substances Directives, households with the Urban Waste Water Treatment Directive). In parallel, other sectorial legislation aimed to protect waters by setting quality criteria for particular uses (bathing with the Bathing Water Directive, food production with the Shellfish Water Directive).

During the last decade, more integrated approaches have been developed to address the issue of unsustainable use of water resources: the Water Framework Directive for the management of inland, transitional and coastal waters using river basin as management units; the Marine Strategy Framework Directive for the management of Europe's regional seas; and a Recommendation on Integrated Coastal Zone Management for coastal areas.

This body of policies is complemented by long-standing agreements between Member States and third countries formalized in regional seas conventions, namely the OSPAR (North Sea), the Helsinki Convention (HELCOM, Baltic Sea), the Barcelona Convention (Mediterranean Sea), and the Bucharest Convention (Black Sea). These cooperative agreements are based on common decisions to protect the marine environment of the regional sea at stake, covering issues ranging from the conservation of marine animals and ecosystem to coastal development and chemical wastes.

A short summary of the scope of current directives and international agreements, and their relevance for coastal zones, is provided in Table 3.1.

Table 3.1 EU legislation and regional conventions regulating water quality in Europe's coastal zones.

Policy	Scope	Relevance for coastal zone
Water Framework Directive (Directive 2000/60/EC)	To achieve good chemical and ecological status in all European surface and ground-waters by 2015	In coastal areas this Directive covers transitional waters and marine waters up to 1 nm from the territorial baseline of Member States for a Good Ecological Status and up to 12 nm for a Good Chemical Status
Floods Directive (Directive 2007/60/EC)	To establish a common framework for assessing and reducing the risk that floods pose to human health, the environment, cultural heritage and economic activity	This Directive covers flooding from the sea in coastal areas
Marine Strategy Framework Directive (Directive 2008/56/EC)	To achieve or maintain Good Environmental Status of the European marine waters by 2020	It deals with coastal waters as defined by Directive 2000/60/EC, their seabed and their subsoil, in so far as particular aspects of the environmental status of the marine environment are not already addressed through that Directive or other Community legislation
Integrated Coastal Zone Management Recommendation (Recommendation 2002/413/EC)	The European Union recommends that Member States adopt a strategic and integrated approach to the management of their coastal zones	No definition of coastal zone/waters is foreseen
Habitats Directive (Directive 1992/43/EEC)	To ensure biodiversity by conserving natural habitats and wild fauna and flora in the territory of the Member States	Several coastal aquatic habitats of community interest are subject to this directive and are included in the Natura 2000 network
Nitrates Directive (Directive 1991/676/EEC)	To reduce and prevent water pollution caused or induced by nitrates from agricultural sources	No direct reference to coastal zones, but the eutrophic state of all water bodies (including coastal waters) must be taken into account in order to initiate/update the designation of vulnerable zones

(*Continued*)

Table 3.1 EU legislation and regional conventions regulating water quality in Europe's coastal zones (*Continued*).

Policy	Scope	Relevance for coastal zone
Discharge of dangerous substances: protection of the aquatic environment (Directive 2006/11/EC). Repealed by the WFD in 2013	To eliminate pollution caused by the discharge of substances in list I of Annex I to the Directive. Also to reduce pollution caused by products in list II of Annex I to the Directive	In the coastal area it applies to 'internal waters', defined as those on the landward side of the baseline from which the breadth of territorial waters is measured, extending, in the case of watercourses, up to the fresh water limit
Urban Waste Water Treatment Directive (Directive 1991/271/EEC) amended by Directive 98/15/EEC	To ensure that all significant discharges of sewage (public services and industrial sectors) undergo treatment before release into surface waters, estuaries or coastal waters	Coastal waters are defined as the waters outside the low-water line or the outer limit of an estuary. Some derogations apply, provided that Member States demonstrate that the discharges do not adversely affect the environment
Bathing Water Directive (Directive 2006/7/EC), repealing by 2014 the old Bathing Water Directive 76/160/EEC	To improve bathing water quality enabling water monitoring and management measures to be improved, and information to be made available to the public	It includes coastal zones as defined in Directive 2000/60/EC
Quality of Shellfish Waters Directive (Directive 2006/113/EC) replacing Council Directive 79/923/EEC. Repealed by the WFD in 2013	Establishes compulsory quality criteria for Member States' shellfish waters	It applies to coastal and brackish waters designated by the Member States as needing protection or improvement in order to support shellfish (bivalve and gasteropod molluscs) life and growth and thus to contribute to the high quality of shellfish products directly edible by man
OSPAR Convention (1992), deriving from the unification of the Oslo (1972) and the Paris (1974) Conventions	It aims to protect and conserve the ecosystems and biological diversity of the North East Atlantic, as well as to prevent and eliminate all kind of pollution	It applies to the 'maritime area', which includes at least the 'internal waters' as defined by Directive 2006/11/EC and the territorial seas
Helsinki Convention (1992), built upon the 1974 Convention	To help reduce pollution in the Baltic Sea and to restore and safeguard its ecological balance	It includes the internal waters (i.e. waters on the landward side of the baseline from which the breadth of the territorial sea is measured up to the landward limit according to the designation by the Contracting Parties) and territorial seas, comprising the water bodies and the seabed
Barcelona Convention (1995), which amends the 1976 version	To reduce pollution in the Mediterranean Sea and protect and improve the marine environment in the area, thereby contributing to its sustainable development	Its revision in 1995 applies to coastal areas as defined by each Contracting Party within its own territory
Bucharest Convention (1992)	To prevent, reduce and control the pollution in the Black Sea in order to protect and preserve the marine environment and to provide legal framework for co-operation and concerted actions to fulfil this obligation	It shall apply to the waters landward of the baselines from which the breadth of the territorial sea is measured and, in the case of freshwater courses, up to the fresh-water limit. Beyond this, the Convention apply to the whole Black Sea proper including the territorial sea and exclusive economic zone of each Contracting Party

Water Framework Directive (2000/60/EC)

The management of European inland, transitional and coastal waters is largely governed by the Water Framework Directive (WFD). The WFD regulates the identification of European waters and their characteristics, on the basis of individual river basin districts (Figure 3.1), and the adoption of management plans and programs of measures with the aim of reaching good chemical and ecological status by 2015 for each body of water. Under its article 14, the directive calls for active involvement of stakeholders and the public in the implementation.

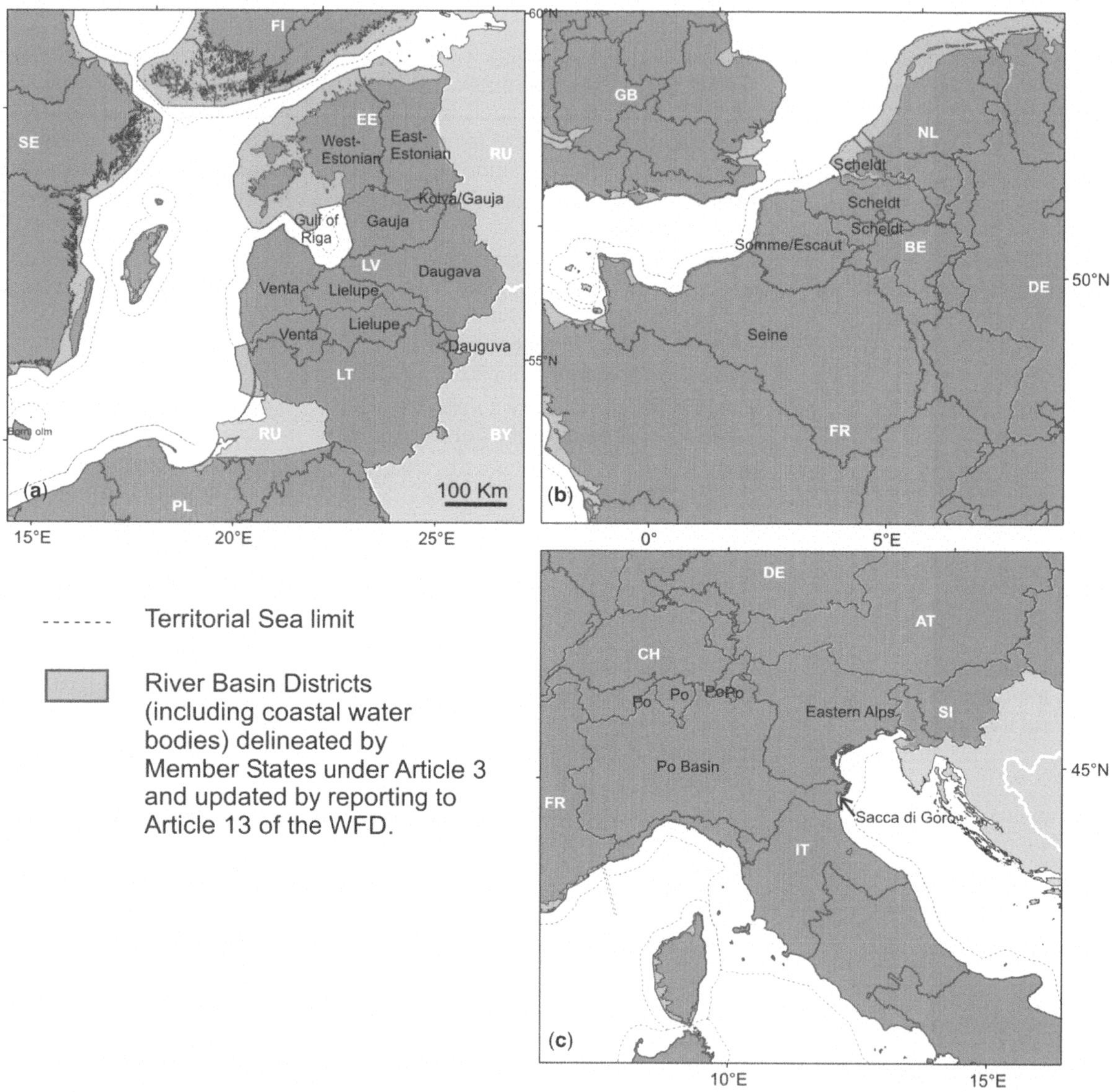

Figure 3.1 Delimitation of River Basins Districts compliant to the WFD around the three AWARE case studies: a) Gulf of Riga, b) Southern North Sea, c) Sacca di Goro. Data from EEA (2011).

Since 2006 the WFD has been amended by several decisions or directives regulating, among others, priority substances or environmental quality standards. New daughter directives regulate particular aspects mentioned in the WFD such as the quality of groundwater or additional quality objectives (e.g. the Groundwater Directive – 2006/118/EC or the Directive on environmental quality standards in the field of water policy – 2008/105/EC). A few existing directives regulating specific issues will be progressively repealed by the WFD. This Directive follows an "adaptive management approach", in that three management cycles are foreseen, with the second and the third one taking into account results and shortcomings from the previous cycle. The first cycle took place between 2000 and 2010. The end of the third cycle marks the final deadline for the achievement of objectives.

The WFD has also competence for transitional waters and coastal zones. Transitional waters are defined as bodies of surface water in the vicinity of river mouths which are partly saline in character as a result of their proximity to coastal waters, but which are substantially influenced by freshwater flows. Coastal waters are defined as surface waters on the landward side of a line every point of which is at a distance of one nautical mile on the seaward side from the nearest point of the baseline from which the breadth of territorial waters is measured, extending where appropriate up to the outer limit of transitional waters. This directive applies to the broader territorial seas only in relation to the chemical status. Figure 3.1 shows some examples of the national delimitations, although some inconsistencies exist between these national delimitations and the definition of coastal waters according to the WFD (Liquete *et al.* 2011).

Floods Directive (2007/60/EC)

The implementation of this Directive must be closely coordinated and synchronised with the implementation of the WFD. The proposed prevention and management measures are organised by river basin districts, as established by the WFD, unless Member States assign specific coastal areas or individual river basins to a different unit of management. The Directive requires the assessment and mapping of the flood risk in river basins, and the design of flood risk management plans.

This Directive also covers floods from the sea in coastal areas, which means that Member States are required to (a) undertake a preliminary flood risk assessment to identify coastal areas at potential risk of flooding; (b) develop flood hazard maps and flood risk maps in coastal areas with relatively high risks of flood damage; and (c) develop flood risk management plans for those zones focused particularly on prevention, protection and preparedness.

Marine Strategy Framework Directive (2008/56/EC)

The Marine Strategy Framework Directive (MSFD) was designed as an extension of the WFD to marine environments, both in its goals and its implementation. Coastal waters must comply with the MSFD for those particular aspects of the status of the marine environment that are not already addressed by the WFD (e.g. noise or litter).

Assessments and plans are coordinated at the Regional Sea level (Baltic Sea, North-East Atlantic, Mediterranean and Black Sea). Within the geographical boundaries of the existing Regional Sea Conventions, cooperation is already taking place between Member States and neighbouring countries sharing the same marine waters.

The Member States must firstly assess the ecological status of their waters and the impact of human activities, and then determine the good environmental status of the waters on the basis of criteria such as biodiversity, the presence of non-indigenous species, stock health, the food chain, eutrophication, changes in hydrographical conditions and concentrations of contaminants, the amount of waste and noise pollution. On the basis of the evaluation of waters, the Member States must define the objectives and indicators to achieve this good environmental status.

Integrated Coastal Zone Management Recommendation (2002/413/EC)

The Integrated Coastal Zone Management (ICZM) represents to date the most far-going attempt to envision coastal landscapes as particular land-sea units for policy development. It is the result of long standing initiatives concerning the state of Europe's coastline, and builds upon the concept of integrated management. ICZM promotes an integrated approach involving all relevant stakeholders, and takes on a long-term and cross-sector perspective in the management of coastal zones. It attempts to balance the needs of development with protection of the very resources that sustain coastal economies, taking into account the public's concern about the deteriorating environmental, socio-economic and cultural state of the European coastline. Coastal zones are not spatially defined in the recommendation and its implementation is based on voluntary cooperation of the Member States.

Habitats Directive (1992/43/EEC)

The cornerstone of Europe's biodiversity policy is the natural Habitats Directive and its ecological network of special protected areas Natura 2000, complemented by the Birds Directive and the new Biodiversity Strategy to 2020. With the Habitats Directive, the EU is seeking to ensure the preservation of biodiversity by conserving key natural habitats and wild fauna and flora in the territory of the Member States. The effect of the Natura 2000 network is enhanced coherence by programs involving monitoring and surveillance, reintroduction of native species, introduction of non-native species, research and education. Many coastal habitats (terrestrial or aquatic) are protected under the Habitats Directive. Member States must take all necessary measures to prevent the deterioration and guarantee the conservation of habitats in designated special areas of conservation.

Nitrates Directive (1991/676/EEC)

The Nitrates Directive is an integral part of the WFD and is designed to protect the Community's waters against nitrates from agricultural sources (diffuse inputs). Member States must identify surface and ground waters polluted or threatened by agricultural nitrogen, with special emphasis on those water bodies where nitrate concentrations exceed 50 mg/l. Member States must then define vulnerable zones (Figure 3.2) which contribute to pollution of each water body. National authorities in turn establish a code of good agricultural practice to be implemented by farmers on a voluntary basis, and set up compulsory action programmes to be implemented by all farmers whose activities fall within the boundaries of designated vulnerable zones. Member States must monitor water quality, applying standardised reference methods to measure the nitrate content.

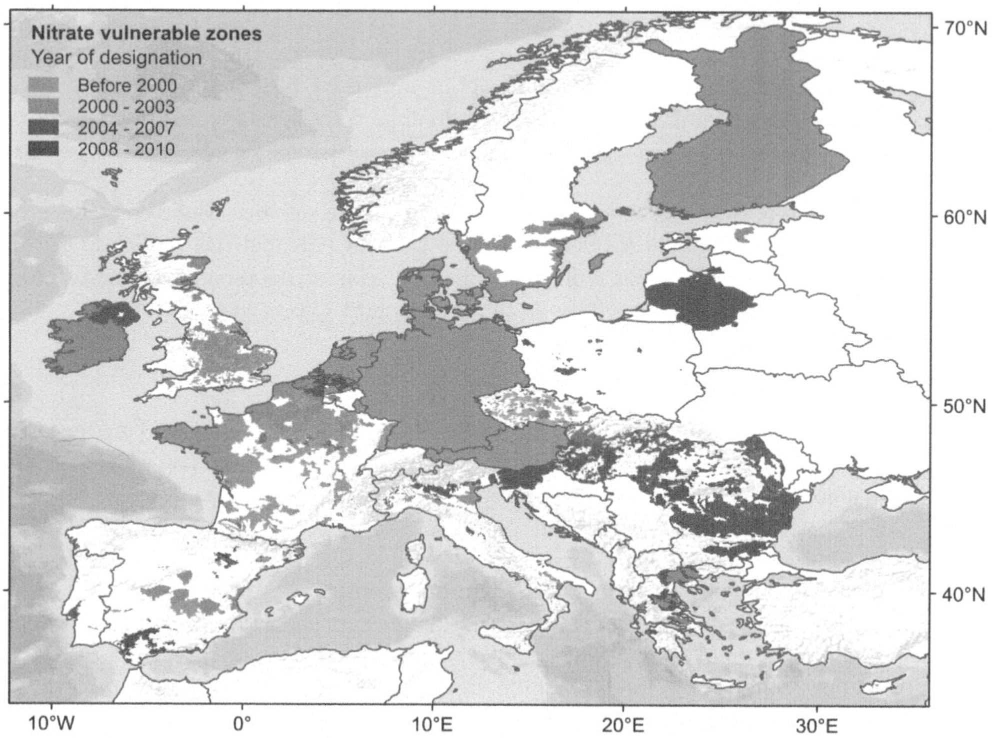

Figure 3.2 Data on designated nitrate vulnerable zones as submitted by Member States for monitoring progress of implementation of the Nitrate Directive. Modified from IES (2009), data kindly provided by V. Juskevicius and K. Bogucarskis.

The coastal zones are not the target of this directive, which aims to protect surface and ground waters by reducing nitrogen inputs from agricultural land. Only the trophic state of coastal zones has to be taken into account for the designation of vulnerable areas. Coastal waters can be heavily affected by these diffuse sources of pollution, as illustrated in the AWARE case studies (see Chapters 6, 7 and 8 of this book). As an example, France was condemned by the EU Court of Justice in 2003 for failing to identify many vulnerable zones, including the Seine Bight (which in turn may lead to identify the whole Seine Basin as a vulnerable zone).

Discharge of dangerous substances directive: protection of the aquatic environment (2006/11/EC)

The regulation of dangerous or hazardous substances in European waters began with Directive 76/464/EEC, now integrated into the WFD and codified as Directive 2006/11/EC on discharge of dangerous substances (DDSD). This Directive links closely to Directive 2008/105/EC on environmental quality standards. It will be repealed by the WFD in 2013.

DDSD presents the rules for protection against, and prevention of, pollution resulting from the discharge of certain substances into the aquatic environment. The directive sets quality objectives and emission limit values for list I substances based on the best available techniques. These limit values are compulsory unless a Member State proves that the quality objectives are being met and consistently maintained. Geographically, DDSD covers the waters as far as the land side of the territorial baseline, which can be located as close as the coastline or as far as several nautical miles offshore, depending on the Member State.

A further improvement and coordination of the water monitoring and pollution control has been established through coordination with Directive 2009/90/EC on technical specifications for chemical analysis and monitoring of water status pursuant to Article 8(3) of WFD. The objective of this last directive is to establish common standards for chemical analysis and monitoring of water, sediment and biota carried out by Member States.

Urban Waste Water Treatment Directive (1991/271/EEC)

Due to their volume, discharges of urban waste water are the second most serious cause of eutrophication, leading in turn to accelerated biodiversity losses, loss of viable drinking water supplies and/or bathing water sites and, thereby, they have important links to public health concerns. The UWWTD regulates urban wastewater collection, treatment and discharge, and wastewater treatment and discharge from the agro-food sector. It required Member States to collect and treat all urban wastewaters by 2005. Newly adjoined Member States (e.g. Estonia and Latvia) are requested to reach the objective by 2015. Specifically, the Directive requires the collection and treatment of waste water in all agglomerations of more than 2000 population equivalents (p.e.) and more advanced treatment for agglomerations of more than 10000 p.e. or in designated sensitive areas.

The latest (6th) Commission Summary analysing the implementation status of this Directive reports that, over 15 years after the coming into force of the UWWTD, the wastewater treatment situation in Europe has improved significantly but the overall goal of the Directive is still not reached: only Austria, Denmark and Germany fully comply, while the remaining Member States fail to meet at least one of the implementation deadlines (on collection, or secondary treatment, or more stringent treatment installations). During the period 1985–2005, the UWWTD has been successful in decreasing nitrogen and phosphorus emissions. However, it is clear that full implementation of the UWWTD constitutes a pre-requisite for meeting the environmental objectives set out in the WFD and in the MSFD.

Bathing Water Directive (2006/7/EC)

Member States have until December 2014 to implement the new Bathing Water Directive (BWD). The Directive's aim is that all bathing waters in the EU must be at least of sufficient quality by the end of the 2015 bathing season. With respect to the old one, the new BWD includes better management with simple sampling and monitoring and is integrated within the body of measures protecting the quality of EU waters through the WFD. The new BWD calls for stricter standards than the old one and, on the basis of recent scientific evidence (Mansilha *et al.* 2009), reduces the burden of the testing protocol.

More relevant for the AWARE project, appropriate monitoring must be also implemented if there is a risk of eutrophication, in which case the authorities must adopt appropriate management measures, provide timely information to the public, and assess possible health risks. The parameters determining water quality in Annex I of the new Directive are stricter for transitional and coastal waters than for inland waters. As an illustration, Figure 3.3 presents the bathing water quality of the season 2010 at coastal sites.

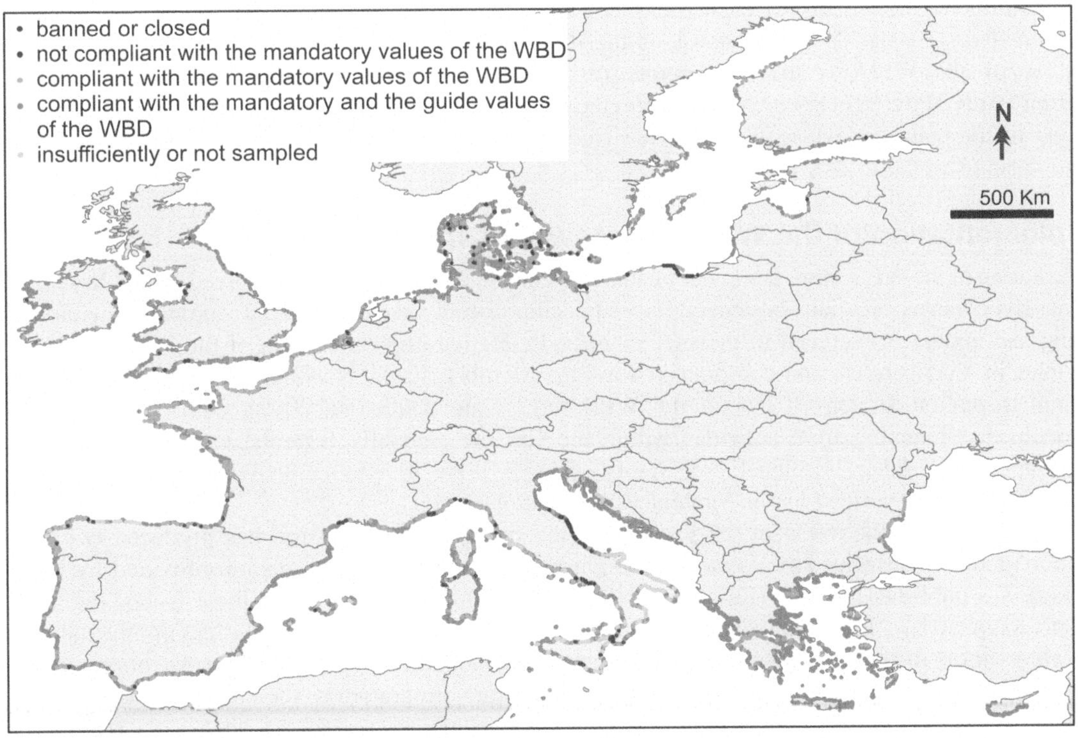

Figure 3.3 Information on bathing water quality as reported by the Member States and Croatia for the 2010 bathing season. In only includes information on transitional and coastal waters. Based on data from ETCW (2011).

Shellfish Water Directive (2006/113/EC)

The Shellfish Water Directive applies to those coastal and brackish waters in which shellfish farming takes place, so as to guarantee the quality of shellfish products intended for human consumption. It is the Member States' responsibility to designate these waters. The Directive establishes a monitoring program, indicating which parameters are to be checked (pH, temperature, salinity and the presence or concentration of certain substances, such as dissolved oxygen, hydrocarbons, metals, organohalogenated substances, etc.), setting indicative and mandatory/limit values, reference methods of analysis and the minimum monitoring/sampling frequency. This Directive will be repealed by the WFD in 2013.

Within the framework of the AWARE project, this Directive is particularly relevant to the case study Sacca di Goro (Italy) in which the main economic sector is clam farming.

Regional Sea Conventions

The OSPAR Convention refers to the "Convention for the Protection of the Marine Environment of the North-East Atlantic"; the Helsinki Convention to the "Convention on the Protection of the Marine Environment of the Baltic Sea Area"; the Barcelona Convention to the "Convention for the Protection of the Marine Environment and the Coastal Region of the Mediterranean"; and the Bucharest Convention to the "Convention on the Protection of the Black Sea Against Pollution". The OSPAR, Helsinki and Bucharest conventions have their own secretariat (OSPAR, HELCOM and Black Sea Commission, respectively) while the Barcelona Convention operates under the UNEP Mediterranean Action Plan.

The four Regional Sea Conventions essentially aim to protect the respective regional seas from marine pollution and eutrophication via intergovernmental cooperation. They usually work with recommendations and decisions on which the signing parties agree and report. The common instruments for these decisions are protocols and, in some cases, specific action plans. Decisions often express the level of reduction of the emission of nutrients or certain pollutants that the signing parties must achieve by a certain deadline. Amongst these conventions, OSPAR and HELCOM in particular have the longest history of cooperation, with active monitoring and assessing of the marine environment, and have become key actors in the management of their respective regional seas.

Coastal waters are within the area of competence of the four conventions. Only the original Barcelona Convention adopted in 1976 excluded internal waters from its text, but it was revised in 1995 to include them, as stated in Table 3.1.

3.3 INSTITUTIONAL FRAMEWORK: NATIONAL IMPLEMENTATION OF EU POLICIES

As illustrated in the previous section, numerous international instruments are in place to regulate the quality of coastal waters. The institutional framework of each Member State for the implementation EU policies varies depending on the specific directive at stake. However, most of the discussions during the AWARE project were focused on the implementation of the WFD as the cornerstone of the EU water policy. Thus, this section summarizes the implementation status of the WFD, which has direct consequences for the management of coastal water quality, with an especial focus in the countries where the AWARE study sites are situated. In addition, a summary is given on the implementation status of the ICZM recommendation.

3.3.1 Implementation of the Water Framework Directive

The implementation of the WFD raises a number of shared technical challenges for all involved actors. In addition, many of the European river basins are trans-boundary, crossing administrative and territorial borders. Therefore a common understanding and approach is crucial to the successful and effective implementation of the WFD. In order to address these challenges in a cooperative and coordinated way, the Member States, Norway and the Commission agreed on a Common Implementation Strategy (CIS) for the WFD (e.g. Water Directors, 2009). Although not legally binding, guidance documents stemming from activities within the CIS are generally regarded as leading the implementation process. To support the latter, technical preparatory work is done in international thematic working groups, and coordinating activities are executed by the Strategic Coordination Group.

After the publication of the WFD in the year 2000, the European Commission has produced two reports on the implementation status of different articles within the Directive based on the information provided by Member States. The first report was published in 2007 (European Commission, 2007b) and addresses three issues: the transposition of the WFD into national law, the set-up of administrative structures (Article 3), and the environmental and economic analysis of river basin districts (Article 5). The second implementation report (European Commission, 2009) was published in 2009 and analyses the establishment of monitoring programmes for surface waters and groundwater (Article 8 and Annex V). Most of the available information, including part of the latest data reported in the river basin management plans (Figure 3.4), can be found in the Water Information System for Europe (WISE, http://water.europa.eu/).

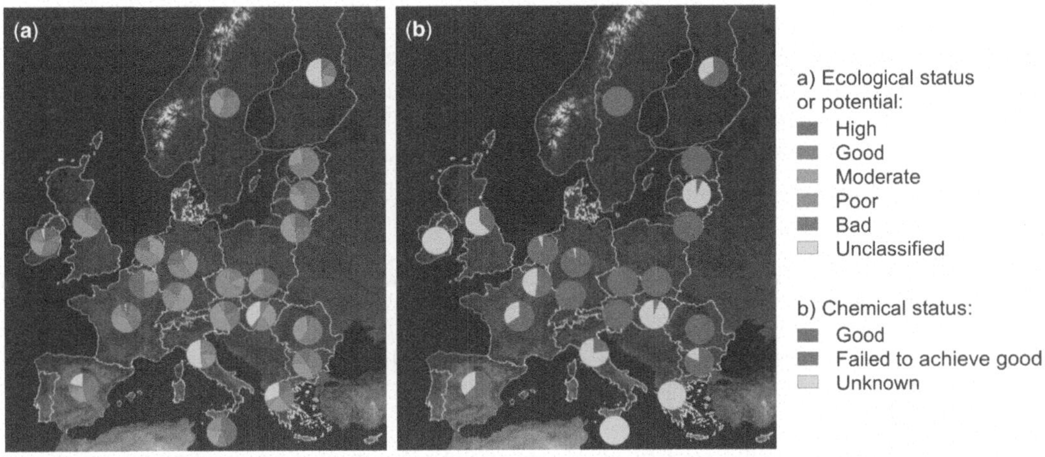

Figure 3.4 Most recent available information on the surface water quality reported in the river basin management plans under the WFD (Article 13) as reported by the Member States. a) Ecological status or ecological potential of surface water bodies. b) Chemical status of surface water bodies. Pie charts show all water bodies by count. Maps are extracted from WISE (2012).

At the time of the first Commission report in 2007, transposition of the directive into national law was largely inadequate. The national administrative structures had been established with only a few but long delays, while most Member States in international river basin district had agreed on coordination arrangements. In general, most countries put considerable effort into the first environmental and economic analysis leading to new relevant information. However, the quality and the level of detail of the reports differed significantly: the largest gaps were related to economic analysis and considerable lag was reported for a number of Member States in incorporating economic instruments into water management systems.

In the report, the Commission assessed the overall reporting performance of Member States based on the time of delivery of the national reports, their clarity and their completeness. Related to the AWARE case study sites, only Estonia is marked above the EU-27 average of reporting performance while Latvia, Belgium, France, and especially Italy are considered below the average standards.

The second implementation report in 2009 highlighted that a large and generally dense network of monitoring stations had been put in place by EU countries, setting the basis of coherent monitoring programmes for the assessment of the status of surface and groundwater. Yet, some gaps were still present in individual river basin districts or individual water categories. Regarding the coastal zone, 12% of the surface waters' monitoring stations (i.e. nearly 7000 stations) were reported as located on transitional or coastal waters. The main concerns at that stage of the implementation were the weak international cooperation in trans-national river basins and the gaps in biological assessment methods. Overall, it was noted that monitoring rules and guidelines were generally followed by Member States but improvement was needed, notably in the specific monitoring of protected areas.

There are at least three biological quality elements required for the assessment of the good ecological status of coastal waters: phytoplankton, other aquatic flora and macroinvertebrates. As regards the coastal waters of the AWARE case study countries, France measures the three biological quality elements in nearly half of its operational monitoring stations, Belgium measures two of the parameters in more than half of its stations, Italy monitors one of the parameters in only 5% of its coastal stations, Estonia has no operational monitoring in coastal waters, and in Latvia the calculation is not possible.

In compliance with the WFD, Member States are required to prepare river basin management plans (RBMP) detailing how the objectives set for the river basin (ecological status, quantitative status, chemical status and protected area objectives) will be reached by 2015. The RBMPs for each river basin district were due to be published at the end 2009, and had to be reported to the Commission by March 2010, together with the results of the public consultations. By December 2011, though, Belgium, Portugal, Spain and Greece had not submitted their RBMP. The Commission is currently evaluating the RBMP that have been timely published.

At the same time, realizing that the objectives of the WFD are far from being timely reached in 2015, the Commission will publish at the end of 2012 the "2012 Blueprint to safeguard EU waters", which will contain a set of policy recommendations aiming to ensure good quality water in sufficient quantities for all legitimate uses. The Blueprint will be based on the results of four ongoing assessments (RBMPs, review of the policy on water scarcity and drought, evolution of water resources, outcome of the fitness check of EU freshwater policy). More information on the current advancement of the Blueprint can be found at: http://ec.europa.eu/environment/water/blueprint/index_en.htm.

Case study Sacca di Goro, Italy

The institutional mapping in the Italian case study is especially complex due to multiple national, regional and local competences split between institutions across different levels and sectors. The Italian authority responsible for the implementation of the WFD in Sacca di Goro is the Directorate General of Environment, Soil and Coast Protection of the Emilia-Romagna Region (Direzione Generale Ambiente, Difesa del Suolo e della Costa – Servizio Tutela e Risanamento della Risorsa Acqua, http://ambiente.regione.emilia-romagna.it/) with the support of other public organizations, in particular:

- Technical support is provided by ARPA (Regional Agency for Environmental Protection), which is charge of monitoring and water quality assessment. Specifically, the Oceanographic Service of ARPA (http://www.arpa.emr.it/daphne/) is responsible for WFD monitoring activities in transitional waters. In Italy, there are currently 21 regional environmental agencies (ARPAs) coordinated by the Institute for Environmental Protection and Research (ISPRA), which acts under the policy guidance of the Italian Ministry for the Environment and the Protection of Land and Sea.
- The Po River Basin Authority (http://www.adbpo.it) has the mission of implementing actions for achieving the WFD goals in the Po River Basin (see Figure 3.1), transitional waters included, and is responsible for the Po river basin management plan. The management plan is performed jointly by the Po River Authority with the regions belonging to the watershed. In this context, the Po River Authority organized several events of public participation with stakeholders, end users, associations of citizens, representatives of municipalities, provinces and regional offices, scientific institutions and technical agencies.
- The Local Agency for Public Health (AUSL) of Ferrara (http://www.ausl.fe.it) is competent for monitoring faecal pollution as indicated in the Bathing and Shellfish directives. On this subject, the license (concession) for shellfish farming is delivered by the regional authority, but the management of farming activities in the lagoon is controlled by the Province of Ferrara (http://www.provincia.fe.it/acquecostiere)
- The Regional Park of the Po Delta (http://www.parcodeltapo.it/) is responsible for the Ramsar sites (Wetlands of International Importance) and is contributing to the implementation of the Habitat and Birds directives through the Natura 2000 network. The application of such directives is also a task of the Nature and Park Conservation Service of the Emilia-Romagna Region.
- The application of the UWWTD is a task of the Administration of the Province of Ferrara, with support from the ARPA for wastewater control (see also the Italian Decree Law No. 152/2006).
- Other decision makers that may regulate the activities (and, therefore, the environmental protection) in Sacca di Goro are the local municipality (http://www.comune.goro.fe.it/), the Local Water Authority (Consorzio di Bonifica, http://www.1bonife.it), the Office of Coastal Waters and Fisheries (Acque Costiere ed Economia Ittica, http://www.provincia.ferrara.it/acquecostiere) that manages in particular the Sacca di Goro lagoon and nearby internal waters, and the Ferrara Optimal Territorial Authority (L'Autorità di Ambito Territoriale Ottimale di Ferrara, http://www.atoferrara.it/) that coordinates the local municipalities to manage in an integrated way the hydrological resources and municipal solid waste.

Case study Gulf of Riga, Latvia and Estonia

The Gulf of Riga is shared by two Member States, Latvia and Estonia. In Latvia, the Ministry of Environment (www.vidm.gov.lv) has overall competence for implementation of the WFD in its four river basin districts. The Latvian Environment, Geology and Meteorology Centre (www.lvgmc.lv) is the competent authority for river basin management planning including coastal and transitional waters as defined by the WFD. Estonia has three river basin districts in which the Ministry of the Environment (www.envir.ee) is the competent authority covering the WFD implementation, also in transitional waters and coastal zones. Nevertheless, the catchment area draining into the Gulf of Riga also includes parts of Lithuania and the non-EU member states Russia and Belarus.

The Latvian Law on Water Management, which integrates key requirements of the WFD in Latvia, came into force on October 2002. It designates four river basin districts (Daugava, Lielupe, Gauja and Venta, see Figure 3.1) in Latvia. Institutional public administration structure in Latvia is made up of various administrative units, boards and agencies belonging to 14 ministries. Every regional agency has a certain area under its supervision, which sometimes is different even within the framework of the same ministry. Within the present State administrative system, in none of the authorities (including the Regional Environmental Boards) the territorial division coincides with the four defined river basin districts. Additionally to the Latvian Law on Water Management, there are two laws (Regional Development Law and Spatial Planning Law) setting three planning levels and regulating development planning in Latvia. Regulations of spatial planning from the Cabinet of Ministers determine the requirements for spatial plans concerning also surface and groundwater.

The main Estonian water legislation is the Water Act, which entered into force in 1994. According to the Water Act the local and regional governments, residents and other interested parties from the relevant river basin shall be involved in the preparation of a water management plan. The additional legal act of a water law not directly related to the WFD is the Estonian Building and Planning Act, which states that people should be informed and involved in the planning process. In this framework, large amount of environmental information is provided through the Estonian Environment Information Centre (http://www.keskkonnainfo.ee/english) and the fifteen county environmental departments.

In 2001 the Ministry of the Environment approved the guidelines for the implementation and establishment of the river basin management plans, setting up the responsible Environmental Services to coordinate and control this implementation. In 2004, the Government of the Republic of Estonia adopted a new regulation regarding the nomination of river basins and sub-river basins, aiming at guaranteeing an optimal and efficiently functioning organisation of water management. The main difference of the new legislation with regard to the previous one is the current division of the national territory into three river basins (West Estonia, East Estonia and Koiva, see Figure 3.1) and eight sub-basins in comparison with only one river basin and nine sub-basins earlier. One of these river basins, the Koiva, is shared with the neighbouring Latvia.

One of the main distinguishing characteristics of the catchment area draining into the Gulf of Riga compared with the other AWARE case studies is that the implementation of sewage treatment facilities is still in a developing stage. This has a direct effect on the quality of coastal waters and on the attainment of the WFD objectives. Sewage treatment is an effective and cost-effective measure for decreasing phosphorous inputs, which are the main cause of eutrophication in the area (Gren & Elofsson, 2008; Bryhn, 2009). After the sixth summary report on the implementation of the UWWTD (see previous section), only 7 out of 71 Latvian agglomerations which were connected to sewage treatment plants complied with the Directive's requirements (GHK *et al.* 2006). HELCOM (2007) estimated that advanced (tertiary) treatment was performed on sewage from 34% of the Estonian population, from 18% of Latvians, and from 0% of Russians and Belarusians. Thus, there is still a great potential in the catchment for reducing phosphorous inputs by means of improved sewage treatment. If 80–90% of the population in the catchment would have their sewage subjected to tertiary treatment, this could decrease phosphorous loadings to the Gulf of Riga by 49% compared to the year 2004 (HELCOM, 2007).

Case study Southern North Sea, Belgium and France

The Southern North Sea case study is focused on three river basins, the Scheldt, the Somme and the Seine (see Figure 3.1), and the coastal zone amongst them. This area is characterized by a complex superimposition of administrative and management authorities, as it spreads over two countries and nine administrative regions.

In France, the competence for the management of the Somme and Seine watersheds is in the hands of regional environmental agencies. In particular, the Seine-Normandie Water Agency (www.eau-seine-normandie.fr) and the Artois Picardie Water Agency (www.eau-artois-picardie.fr) are responsible for the Seine and the Somme/Scheldt, respectively. The Seine-Normandie Water Agency is in charge of the implementation of all decisions pertaining to water management, including the WFD requirements. It acts as the executive organism of the Basin Committee, which is formed by representatives of the French State, the regions and the local communities as well as of members of the civil society. The Basin Committee developed, in line with the prescription of the WFD, a general Development Scheme for Water Management (SDAGE) as well as more local measure programmes (SAGE) in order to achieve the objectives of the WFD.

The Scheldt river basin in Belgium is crossing the three different regions of the country which have competence over the environment and for implementation of the WFD. These are: the Brussels Institute for Management of the Environment (http://www.ibgebim.be/index.htm) for the Brussels Capital Region; the Directorate General for Waterways (voies-hydrauliques.wallonie.be) and the Directorate General for Natural Resources and Environment (environnement.wallonie.be) for Wallonia; and the Coordination Committee Integrated Water Policy (www.ciwvlaanderen.be) for Flanders. This latter authority is competent for the Belgian transitional waters, while the Federal Government takes responsibility over the coastal zone.

The Artois-Picardie Water Agency is also involved, with the Walloon, Flemish and Brussels Region, in the management of the Somme/Scheldt watersheds. The works to be carried out in the frame of water management (e.g. wastewater treatment plants) have to be owned and managed by other organisms different from the water agencies. Some examples of such organisms are the SIAAP (Syndicat Interdépartemental pour l'Assainissement de l'Agglomération Parisienne) responsible for Paris wastewater treatment, the IBGE (Institut Bruxellois pour la Gestion de l'Environnement) for Brusels, the VMM (Vlaams Milieumaatschappij) in Flanders, and the SPGE (Societé Publique de Gestion de l'Eau) in Wallonia.

Owing to the efforts devoted to urban wastewater treatment by local and governmental authorities, the point sources of nutrient to surface water are now about to be under control. The requirements of the WFD in that matter and the necessary investments are clear and addressed. On the other hand, nitrogen coming from diffuse agricultural sources is still a great concern in the area, both in terms of alteration of drinking water resources and of coastal eutrophication problems. During the last decades, the Seine and the Somme basins specialized in cereal production relying on synthetic fertilization, while the Scheldt basin shifted to intensive animal farming. Both synthetic fertilizers in crops areas and manure excess in livestock farming areas result in an increase of diffuse sources of nutrients to surface and groundwater. Even if the Common Agricultural Policy after its reform of 1999 (oriented towards a more multi-functional agriculture and agro-environmental measures), the Nitrate Directive and the WFD aim at reducing nitrate inputs in ground and surface water due to farming activities, the imposed agro-environmental measures in the area seem to have only limited effectiveness (see Chapter 7 of this book).

3.3.2 Implementation of the ICZM recommendation

Rupprecht Consult and International Ocean Institute (2006) evaluated the implementation of the EU ICZM recommendation across Europe (Figure 3.5) and identified many suggestions for improvement. Overall, 18 of the 24 coastal Member States and Accession Countries had officially reported on the implementation of the EU ICZM Recommendation by mid-June 2006. National contact points for the implementation of this Recommendation are listed in the report. For the AWARE case study countries the reporting institution are: the Federal Public Service of Health – Directorate General of the Environment – Marine Environment Service (Belgium), the Inter-ministerial Delegation to the Development and Regional Competitiveness (France), the Ministry of Regional Development and Local Governments – Spatial Planning Department (Latvia), and in Italy it is unclear.

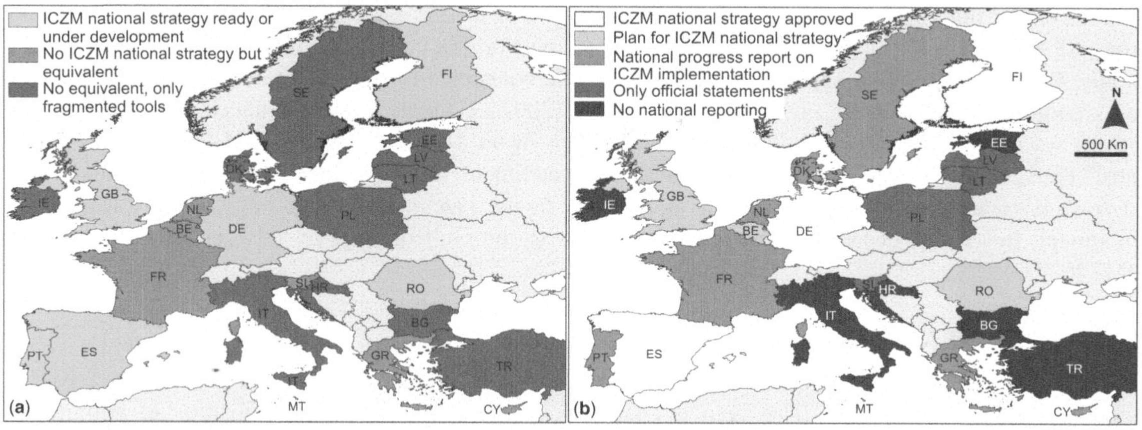

Figure 3.5 Status of the implementation of the ICZM Recommendation (2002/413/EC) in EU Member States and Candidate Countries in 2006. a) Level of implementation in different coastal states. b) National reporting activities from the same countries. Data from Rupprecht Consult and International Ocean Institute (2006).

As a summary, in the 24 EU coastal Member States and Accession/Candidate Countries that were analysed, the status of policy implementation was as follows:

- No country had implemented an ICZM National Strategy as indicated by the EU ICZM EU Recommendation.
- In 7 countries the implementation of an ICZM National Strategy was still pending.
- In 6 further countries documents considered as equivalent to an ICZM National Strategy had been developed, or coastal zone management strategies had become (or planned to become) an integral part of its spatial planning processes.
- In 11 countries no ICZM equivalent policies were in advanced stage of preparation, only fragmented tools were in place to address coastal issues.

One of the main issues found is that ICZM is primarily being implemented as a national strategy with scarce cooperation at the pan-European or regional sea level. The development of this regional dimension of ICZM would have many practically and strategically positive impacts.

The revision of the ICZM national implementation reports highlights that integrated approaches to manage the coastal zone had been only partially implemented and they usually applied to case study basis. Nevertheless, the few available examples of implemented ICZM projects showed that there are major and measurable advantages in areas with ICZM. However, due to particular historic contexts in the way different countries develop and apply procedures and processes in coastal areas, a real holistic and participatory management approach in the coastal zone will take time (Rupprecht Consult and International Ocean Institute, 2006).

3.4 ANALYSIS OF THE POLICY FRAMEWORK. INSIGHTS FROM THE AWARE EXPERIENCE

In the coastal area, most of the physical, chemical and biological processes are affected directly by land inputs and human activities, mainly through rivers courses. In this chapter the relatively long history of EU water policies and international conventions is presented, addressing some important pressures in coastal zones, like pollution by waste water discharges, nutrients from agriculture, or industrial emissions. However, the state of compliance and implementation of the various directives is irregular across Europe and the impacts of man-made pressures are far from controlled. This is mainly due to the historical, political and economic reasons within each country, and to the priorities set in addressing the pressures on coastal waters. There are issues related to who should pay for the implementation of the directives in the respective countries. During the AWARE citizens panel meetings it was observed that the perceptions of citizens from different case studies differed on who should pay for the implementation of the WFD and the MSFD. In some cases, the citizens felt that the current policy and institutional framework is too cryptic and complicated, and that they were not well informed by scientists or EU institutions on these issues; in others, it was felt that the conflict between different stakeholders or actors and economic interests makes the implementation of the directives difficult. Thus, that there is a strong need for strengthening the science-policy linkage and also the involvement of citizens and stakeholders in the development and implementation of the directives in the future. Citizens in general are interested to take part and also to share the responsibility in maintaining a good quality status of the coastal water bodies by increasing awareness about the problems and remedial measures.

Still at present, excessive nutrient inputs from diffuse sources of pollution continue to pose a major threat to the European coastal waters (EEA, 2010). After over 15 years from its publication, the implementation of the Nitrates Directive was still incomplete, leading to several infringement procedures mainly related to the designation of nitrate vulnerable zones and the non-conformity of action programmes (Jack 2006; European Commission 2007a). The reinforcement of the existing environmental regulations was dealt with at great length during the AWARE meetings and was one of the main issues raised by the AWARE citizens' panel as mentioned before.

Given the broad range of policy tools currently in place, one would feel that the European coastal waters should be well protected. Still, current state of achievements indicates that there is room for improvement. To start with, the WFD and the MSFD should be fully integrated to promote a continuous assessment and management from land to sea (Borja *et al.* 2010; Liquete *et al.* 2011). In this framework, the largest gaps to be covered relate to policies regulating particular economic sectors, like agriculture (European Commission, 2010b) or energy (European Commission, 2010a), that act as drivers for water management choices. The cross compliance of those policies with the WFD is foreseen but still not fully applied. For example, the new strategy for a Resource Efficient Europe (European Commission, 2011) anticipates the revision of environmentally harmful subsidies. The lack of cross compliance, especially amongst the Common Agricultural Policy and the WFD, was quickly perceived by the AWARE citizens' panel and represented one of the pillars of the final Citizens Declaration. Amongst the potential alternatives, Lindahl and Kollberg (2009) proposed cost-effective measures to reduce eutrophication in coastal waters based on an extension of the EU agro-environmental aid program into the coastal zone.

The WFD takes on an innovative and ambitious ecosystem-based approach to manage water resources focusing on river basins, while including at the same time the adjacent coastal area. It represents a challenge and an opportunity to protect the coastal ecosystems and develop real coastal zone management (Borja, 2005; de Jonge *et al.* 2006). One of the most difficult tasks within this Directive is the development of coherent and comparable national monitoring programmes. In addition to the limited biological quality measures mentioned before as one the weak points in the implementation of the WFD, the development of consistent cause-effect studies remains one of the main issues in the assessment of the status of surface waters in general and coastal waters in particular (de Jonge *et al.* 2006; Birk *et al.* 2012). After analysing and comparing existing monitoring programmes, several researchers have proposed practical strategies to monitor coastal and transitional waters more efficiently under the WFD (Ferreira *et al.* 2007) or the MSFD (Zampoukas *et al.* 2012).

Due to the multiple uses and actors interested in coastal areas, there is an increasing need of practical sea-use planning in Europe (European Commission 2007c). Several European countries have taken global leadership in implementing marine spatial planning initiatives, as a consequence of having to develop national ICZM strategies (Douvere and Ehler, 2009). Some scenarios on policy options relevant to ICZM in Europe estimate that a "no-control" policy option would generate 8% more built-up areas in coastal zones than a policy option geared toward sustainability, with a consequent increase in pollution, impacts on the water cycle and vulnerability of coastal areas to erosion and flooding (Lavalle *et al.* 2011). The development of marine spatial planning would encourage the consolidation of coastal and marine information to support policy implementation and promote ecosystem-based management of coastal and marine areas (Meiner, 2010). The further step toward a holistic approach to environment, already mentioned in legislation like WFD or ICZM but very rarely implemented, is to assess the entire environmental-human systems, as proposed by Bigagli (2012) for the marine environment.

Any new policy, directive or recommendation needs to be supported by proper institutional framework, funding and capacity building within each country. Given the current critical economic situation within the EU, some of the Member States will not prioritize investing in the implementation of the various directives. However, the costs of inaction, both in terms of resources availability and deterioration of environment and biodiversity, have been evaluated as greatly surpassing the burden of investments required by current policies (see, for example, Braat & ten Brink, 2008 or Moss, 2012). The balancing of the larger economic and environmental interests will have to be carefully considered and addresses at the EU level. Active involvement of citizen panels, a methodology developed in AWARE, could partially address the issue by putting more pressure on national governments and mobilizing the civil society: after all, the ultimate interests of civil society lay in a healthier coastal environment for recreation, fisheries and other services.

3.5 CONCLUSIONS

Developing a participatory process, as demonstrated in the AWARE project, is a resources-consuming operation (time, money) and requires a careful review of the policy and institutional framework. This chapter summarizes the key environmental policies relevant to the protection and management of European coastal waters, notably the Water Framework Directive (WFD). The national authorities responsible for the implementation of the WFD and the Integrated Coastal Zone Management (ICZM) in the AWARE case studies (Gulf of Riga, Southern North Sea and Sacca di Goro) are also included in this compilation. They were contacted during the AWARE project and asked how they perceived the challenges related to the implementation of the various directives for protection of coastal waters. In all three case studies, the cross compliance of the environmental legislation with other crucial sectors, like the Common Agricultural Policy, was highlighted as essential for consistency in the process. Invited speakers and stakeholders were able to express their opinions on the current policy and institutional framework during the AWARE process.

The WFD represents the Europe's main response to the challenges raised by poor water quality and uncoordinated management strategies. The directive sets up an integrated framework aiming to restore aquatic ecosystems to good ecological and chemical status by engaging all stakeholders involved, including the public or end users. This links to one of the AWARE project mottos: "In getting our waters clean, the role of citizens and citizens' groups is crucial". It is also essential to secure funding and to coordinate the various sectors for fulfilment of these objectives. Main gaps regarding the implementation of the WFD are the quality and the level of detail of certain national reports; important data gaps in socio-economic analyses and in biological assessments; weak international cooperation in trans-national river basins; comparability of data; and specific monitoring in protected areas.

The WFD is complemented by the Floods Directive and the Marine Strategy Framework Directive (MSFD). Other supporting EU directives relating to or regulating particular aspects of water quality in coastal zones are: the Nitrates Directive, the "Discharge of dangerous substances: protection of the aquatic environment" Directive, the Urban Waste Water Treatment Directive, the Bathing Water Directive, and the Quality of Shellfish Waters Directive.

Apart from these directives, the EU recommendation on ICZM has the potential to foster the achievement of the objectives of the WFD and the MSFD in coastal zones. However, the recommendation is based on voluntary cooperation of the Member States and has no consequences in case of non-compliance. For instance, by 2006 no country had implemented a National Strategy as prompted by the recommendation. Cooperation amongst Member States and between those and third countries is built further on the existing Regional Sea Conventions: OSPAR, Helsinki, Barcelona and Bucharest conventions. A short review of all these legal instruments together with an analysis of their relevance for coastal zones is provided in this chapter.

The implementation of these environmental regulations has so far not always been timely and/or thorough as required for full compliance, thus giving rise in some cases to infringement procedures. Stricter implementation and control by EU and national institutions was claimed several times during the AWARE process, as reflected in the final Citizens' Declaration.

REFERENCES

Barcelona Convention (1995). Convention for the Protection of the Marine Environment and the Coastal Region of the Mediterranean, adopted in 1976, in force from 1978, revised in June 1995, revised version not yet in force. Available at: http://www.unepmap.org/. (accessed 12 March 2012).

Bigagli E. (2012). Sustainable development and the marine environment. Data and information requirements for an integrated assessment of the marine environment. *Ocean & Coastal Management*, submitted.

Birk S., Bonne W., Borja A., Brucet S., Courrat A., Poikane S., Solimini A., van de Bund W., Zampoukas N. and Hering D. (2012). Three hundred ways to assess Europe's surface waters: an almost complete overview of biological methods to implement the Water Framework Directive. *Ecological Indicators*, **18**, 31–41.

Borja A. (2005). The European water framework directive: A challenge for nearshore, coastal and continental shelf research. *Continental Shelf Research*, **25**, 1768–1783.

Borja A., Elliott M., Carstensen J., Heiskanen A.-S. and Van De Bund W. (2010). Marine management – towards an integrated implementation of the European Marine Strategy Framework and the Water Framework Directives. *Marine Pollution Bulletin*, **60**, 2175–2186.

Braat L. and ten Brink P. (eds.) (2008). The Cost of Policy Inaction: The Case of Not Meeting the 2010 Biodiversity Target. Alterra, Alterra-rapport 1718, Wageningen, p. 312.

Bryhn A. C. (2009). Sustainable phosphorus loadings from effective and cost-effective phosphorus management around the Baltic Sea. *PLoS ONE*, **4**, e5417.

Bucharest Convention (1992). Convention on the Protection of the Black Sea Against Pollution, adopted in 1992, in force from 1994. Available at: http://www.blacksea-commission.org/main.asp. (accessed 12 March 2012).

de Jonge V. N., Elliott M. and Brauer V. S. (2006). Marine monitoring: its shortcomings and mismatch with the EU Water Framework Directive's objectives. *Marine Pollution Bulletin*, **53**(1–4), 5–19.

Directive 1991/271/EEC (1991). Council Directive 91/271/EEC of 21 May 1991 concerning urban waste-water treatment. OJ, L135, 30.5.1991, p. 40.

Directive 1991/676/EEC (1991). Council Directive of 12 December 1991 concerning the protection of waters against pollution caused by nitrates from agricultural sources. OJ, L375, 31.12.1991, 1.

Directive 1992/43/EEC (1992). Council Directive 92/43/EEC of 21 May 1992 on the conservation of natural habitats and of wild fauna and flora. OJ, L206, 22.7.1992, 7.

Directive 2000/60/EC (2000). Directive 2000/60/EC of the European Parliament and of the Council of 23 October 2000 establishing a framework for Community action in the field of water policy. OJ, L327, 22.12.2000, 1.

Directive 2006/11/EC (2006). Directive 2006/11/EC of the European Parliament and of the Council of 15 February 2006 on pollution caused by certain dangerous substances discharged into the aquatic environment of the Community. OJ, L64, 4.3.2006, 52.

Directive 2006/113/EC (2006). Directive 2006/113/EC of the European Parliament and of the Council of 12 December 2006 on the quality required of shellfish waters. OJ, L376, 27.12.2006, 14.

Directive 2006/7/EC (2006). Directive 2006/7/EC of the European Parliament and of the Council of 15 February 2006 concerning the management of bathing water quality and repealing Directive 76/160/EEC. OJ, L64, 4.3.2006, 37.

Directive 2007/60/EC (2007). Directive 2007/60/EC of the European Parliament and of the Council of 23 October 2007 on the assessment and management of flood risks. OJ, L288, 6.11.2007, 27.

Directive 2008/56/EC (2008). Directive 2008/56/EC of the European Parliament and of the Council of 17 June 2008 establishing a framework for community action in the field of environment al policy (Marine Strategy Framework Directive). OJ, L164, 25.06.2008, 19.

Douvere F. and Ehler C. (2009). New perspectives on sea use management: initial findings from European experience with marine spatial planning. *Journal for Environmental Management* **90**, 77–88.

EEA (2010). 10 Messages for 2010 – Coastal Ecosystems. European Environment Agency, Office for Official Publications of the European Union, Luxembourg, 17 pp.

EEA (2011). River Basin Districts (RBDs) version 1.4 (06/2011). Water Information System for Europe, European Environment Agency, available at: http://www.eea.europa.eu/data-and-maps/data/wise-river-basin-districts-rbds-1. (accessed 12 March 2012).

ETCW (2011). Bathing Water Directive, Status 1990–2010. National submissions to the Bathing Water Directive, database June 2011. The European Topic Centre on Water, available at: http://www.eea.europa.eu/data-and-maps/data/bathing-water-directive-status-of-bathing-water-3. (accessed 12 March 2012).

European Commission (2007a). Report from the Commission to the Council and the European Parliament on Implementation of Council Directive 91/676/EEC Concerning the Protection of Waters Against Pollution caused by Nitrates from Agricultural Sources for the Period 2000–2003. COM(2007) 120 final, Brussels, 19.3.2007.

European Commission (2007b). Communication from the Commission to the European Parliament and the Council: Towards Sustainable Water Management in the European Union–First Stage in the Implementation of the Water Framework Directive 2000/60/EC. COM(2007) 128 final, Brussels, 22.3.2007.

European Commission (2007c). Communication from the Commission: Report to the European Parliament and the Council: An Evaluation of Integrated Coastal Zone Management (ICZM) in Europe. COM (2007) 308 final, Brussels, 7.6.2007.

European Commission (2009). Report from the Commission to the European Parliament and the Council in Accordance with article 18.3 of the Water Framework Directive 2000/60/EC on Programmes for Monitoring of Water Status. COM(2009) 156 final, Brussels, 1.4.2009.

European Commission (2010a). Communication from the Commission to the European Parliament, the Council, the European Economic and Social Committee and the Committee of the Regions: Energy 2020: A Strategy for Competitive, Sustainable and Secure Energy. COM(2010) 639 final, Brussels, 10.11.2010.

European Commission (2010b). Communication from the Commission to the European Parliament, the Council, the European Economic and Social Committee and the Committee of the Regions: The CAP towards 2020: Meeting the Food, Natural Resources and Territorial Challenges of the Future. COM(2010) 672 final, Brussels, 18.11.2010.

European Commission (2011). Communication from the Commission to the European Parliament, the Council, the European Economic and Social Committee and the Committee of the Regions: Roadmap to a Resource Efficient Europe. COM(2011) 571 final, Brussels, 20.9.2011.

Eurostat (2011). Coastal region statistics – Statistics Explained. Available at: http://epp.eurostat.ec.europa.eu/statistics_explained/index.php/Main_Page. (accessed 12 March 2012).

Ferreira J. G., Vale C., Soares C. V., Salas F., Stacey P. E., Bricker S. B., Silva M. C. and Marques J. C. (2007). Monitoring of coastal and transitional waters under the E.U. Water Framework Directive. *Environmental Monitoring and Assessment*, **135**, 195–216.

GHK, ECOLAS, IEEP and CE (2006). Strategic Evaluation of Environment and Risk Prevention Under Structural and Cohesion Funds for the Period 2007–2013 – National Evaluation Report for Latvia – Executive Summary. Directorate General Regional Policy, Brussels, p. 17. Available at: http://ec.europa.eu/regional_policy/sources/docgener/evaluation/pdf/evalstrat_env/lv_exec.pdf. (accessed 12 March 2012).

Gren I.-M. and Elofsson K. (2008). Costs and Benefits from Nutrient Reductions to the Baltic Sea. Swedish EPA Report 5877, Stockholm, p. 67.

HELCOM (2007). Towards a Baltic Sea unaffected by eutrophication HELCOM Overview 2007. HELCOM, Krakow, p. 35. Available at: http://www.helcom.fi/stc/files/Krakow2007/Eutrophication_MM2007.pdf. (accessed 12 March 2012).

Helsinki Convention (1992). Convention on the Protection of the Marine Environment of the Baltic Sea Area, adopted in 1974, in force from 1980, revised in 1992, in force from 2000. Available at: http://www.helcom.fi/home/en_GB/welcome/. (accessed 12 March 2012).

IES (2009). FATE and Impact of Pollutants in Terrestrial and Aquatic Ecosystems, Nitrate Vulnerable Zones Viewer (ELISA). European Commission – Join Research Centre – Institute for Environment and Sustainability, available at: http://fate.jrc.ec.europa.eu/elisa-viewer. (accessed 12 March 2012).

Jack B. (2006). Tackling eutrophication: the implications of a precautionary approach. *European Environmental Law Review*, **15**(12), 354–368.

Lavalle C., Rocha Gomes C., Baranzelli C. and Batista e Silva F. (2011). Coastal Zones – Policy alternatives impacts on European Coastal Zones 2000–2050. EUR 24792, Publications Office of the European Union, Luxembourg, p. 104. Available at: http://publications.jrc.ec.europa.eu/repository/. (accessed 12 March 2012).

Lindahl O. and Kollberg S. (2009). Can the EU agri-environmental aid program be extended into the coastal zone to combat eutrophication? *Hydrobiologia*, **629**, 59–64.

Liquete C., Somma F. and Maes J. (2011). A clear delimitation of coastal waters facing the EU environmental legislation: from the Water Framework Directive to the Marine Strategy Framework Directive. *Environmental Science & Policy*, **14**(4), 432–444.

Mansilha C. R., Coelho C. A., Heitor A. M., Amado J., Martins J. P. and Gameiro P. (2009). Bathing waters: new directive, new standards, new quality approach. *Marine Pollution Bulletin*, **58**, 1562–1565.

Martínez M. L., Intralawan A., Vázquez G., Pérez-Maqueo O., Sutton P. and Landgrave R. (2007). The coasts of our world: ecological, economic and social importance. *Ecological Economics*, **63**(2–3), 254–272.

Meiner A. (2010). Integrated maritime policy for the European Union – consolidating coastal and marine information to support maritime spatial planning. *Journal of Coastal Conservation*, **14**, 1–11.

Moss J. (2012). Water and the economic crisis: Back to basics. OECD Observer. Available at: http://www.oecdobserver.org/news/fullstory.php/aid/2845/Water_and_the_economic_crisis.html. (accessed 12 March 2012).

OSPAR Convention (1992). The Convention for the Protection of the Marine Environment of the North-East Atlantic – Oslo and Paris conventions, adopted in 1974, revised and combined into OSPAR Convention in 1992, in force from 1998. Available at: http://www.ospar.org/. (accessed 12 March 2012).

Recommendation 2002/413/EC (2002). Recommendation 2002/413/EC of the European Parliament and of the Council of 30 May 2002 concerning the implementation of Integrated Coastal Zone Management in Europe. OJ, L148, 06.06.2002, 24.

Rupprecht Consult and International Ocean Institute (2006). Evaluation of Integrated Coastal Zone Management (ICZM) in Europe – Final Report. p. 360. Available at: http://ec.europa.eu/environment/iczm/pdf/evaluation_iczm_report.pdf. (accessed 12 March 2012).

Water Directors (2009). Common Implementation Strategy for the Water Framework Directive (2000/60/EC): Work Programme 2010–2012. Available at: http://circa.europa.eu/Public/irc/env/wfd/library?l=/framework_directive/implementation_documents/final_2010-2012/_EN_1.0_&a=d. (accessed 12 March 2012).

WISE (2012). WFD surface water viewer, Water Information System for Europe. Available at: http://www.eea.europa.eu/themes/water/interactive/soe-wfd/wfd-surface-water-viewer. (accessed 12 March 2012).

Zampoukas N., Piha H., Bigagli E., Hoepffner N., Hanke G. and Cardoso A. C. (2012). Monitoring for the Marine Strategy Framework Directive: Requirements and Options. EUR 25187, Publications Office of the European Union, Luxembourg, p. 36. Available at: http://publications.jrc.ec.europa.eu/repository/. (accessed 12 March 2012).

Chapter 4

Building a new Science-Citizens-Policy interface: theoretical foundations

Carlo Sessa

AWARE Project Coordinator, Institute of Studies for the Integration of Systems, Italy

4.1 INTRODUCTION: FOSTERING CITIZENS PARTICIPATION

The AWARE project adopted a pragmatic approach to develop and test a knowledge brokerage process engaging scientists, citizens and decision makers. This approach is mostly based on the citizens conference methodology, originally developed by the Danish Board of Technology and then applied more broadly. This pragmatic approach is extensively described in Chapter 5.

In this Chapter, the focus will be on theoretical foundations underpinning the citizens conference methodology. These can be traced back to two classical developments of political philosophy, namely John Dewey's analysis of the public and its problems and John Rawls' theory of social justice – which justify the extension of the Science-Policy Interface (SPI) concept to the broader concept of Science-Citizens-Policy Interface (SCPI), adopted as a new paradigm in the AWARE method. This analysis of theoretical foundations is needed to show that the methodology tested in the AWARE project is robustly based on sound political concepts.

The nature and scope of democratic politics in modern societies has been a subject of academic and political debate since the early days. A key issue has always been the extent to which it is possible to sustain active and dynamic democratic practices in a large and complex multi-layer State regime. Already in the Federalist Papers, the founding fathers of American democracy Madison, Hamilton and Jay writing under the pseudonym "Publius", debated the importance of federalism for a large, multi-lingual, heterogeneous and divided country like the United States at the time. One of their main arguments was that federalism was a facilitator – and not an inhibitor – of democracy because only federalism with its many levels of government could ensure that there were enough checks and balances on central government but also on powerful interest lobbies in a large and complex system. A few centuries later, the European Union finds itself facing similar, even if more complex, challenges as it seeks to advance beyond economic integration towards political integration within a multi-level governance framework.

Whether the European Union will evolve into a full federal state remains an open question (as this could imply a significant loss of national sovereignty it is evidently not favoured by most national political elites). But even if the future of the European Union lies rather in inter-governmental relations, which delineates a regulatory model of loose coordination (Majone, 1996) rather than a political model of institutional convergence, one thing is certain: that in a multi-level government system it is no longer possible to rely alone on representative democratic institutions (such as the parliament) and procedures (such as voting) to gather citizen input into decision-making. A complex democratic polity requires a multi-faceted *public sphere* (Habermas, 1991, 1998), especially to build a larger and deeper societal consensus on grand sustainability challenges, and this, in turn, calls for more stakeholders' involvement in decision-making (hence governance rather than government – see EC 2002) as well as more latitude for deliberative processes (Schmitter, 2000; Giorgi *et al.* 2006).

Unlike representative democracy, which is conceptually based on the model of delegation, deliberative democracy calls for active citizenship (Barber, 1984; Held, 1996). Active citizens are eager to organize themselves within communities or social movements, and they endeavour to impact on policy-making. However, a larger arena of ordinary citizens may be called into take part in citizen deliberative forums organized by national or, principally, local governments for tapping on citizen views on specific policy proposals. Policy forums, insofar as they are organized by public authorities, are a form of top-down citizens participation, different from bottom-up active citizenship (as in community or social movements).

Both forms of participation, but especially bottom-up community or social movements, are often criticized for their single-issue orientation: their objective is to bring change in one policy area – and often within a strictly delineated sub-component – and this goal is pursued without due consideration to other effects or other groups of users. Top-down citizens participation is further criticized for being more prone to manipulation given its dependency on public authorities or "the policy-maker" for infrastructure and funding. Even though these are dangers that must be borne in mind, it is equally important not to fall into the trap of magnifying them beyond proportion (Giorgi *et al.* 2006). Within the realm of top-down deliberative politics much, in fact, depends on the organizational format and procedures followed. The form of deliberative processes is thought to influence the content (Move Together, 2009).

Citizens participation experiences often differ in terms of their objectives (why is citizens participation being sought? What will the final output be?), design/methodology (who will participate and how will participants be selected? What instrument will be implemented to achieve the objective?), as well as scope (how many citizens should participate? How long and how often will meetings take place? What resources are available?). Despite the various forms citizens participation can take, the underlying purpose should be the intensification of effective participation in and influence of ordinary citizens on policy development and formulation, as their lives and communities are more or less directly impacted by public policies. Ideally citizens participation should also be experienced as "empowering", whereby citizens emerge from the participatory experience with greater knowledge and the feeling that they wish to continue to actively contribute to policy-making in their role as citizens (Fung *et al.* 2003).

Despite the enthusiasm for promoting democracy, most of the calls for more democracy and citizens participation occur at a time when we are witnessing the disturbing decline of democratic practices in the Western democracies. Many factors explain the wane of the citizen's role. Among the most important is the social and technical complexity of modern societies (Fisher, 2005). What are the possibilities, many ask, of the ordinary (i.e., non expert) citizen deliberating intelligently on the policy issues confronting the decision makers of such societies? Are not these issues better addressed by the professional experts? What evidence supports any contention that citizens can effectively participate in helping to make the complex decisions facing contemporary policy makers?

Hard evidence demonstrates indeed that the ordinary citizen is capable of a great deal more participation than generally recognized or acknowledged (Fisher, 2005). Citizens participation is not to be seen as a magic cure-all for economic, social and environmental problems, nor is deliberation meant to direct attention away from questions of interest and power which concretely contribute to shape policy making. But it does hold out the possibility of bringing forth new knowledge and ideas capable of creating and legitimating new interests, reshaping our understanding of existing interests, and, in the process, influencing the political pathways along which power and interest travel. Deliberations of ordinary citizens, if appropriately framed, can help not only in searching solutions to pressing environmental and social problems, but they can also contribute a kind of knowledge – in particular, local knowledge – that the professional expert requires but could not easily access.

Challenging the scientific expert's methodological emphasis on "generalizable knowledge", post-positivist theory underscores the importance of bringing in the local contextual knowledge of the ordinary citizen. To be usable, knowledge has to be applied to a particular situation or context. In this sense, the case for participation is seen to be as much grounded in epistemology as in democratic politics. Appropriately framed citizens participation might best be thought as a way to solve a translation problem: how do we translate abstract propositions to make them fit to particular contexts?

Broadly speaking, therefore, participation contributes to three important goals. First of all, citizens participation and its normative rationale, deliberation, give meaning to democracy. If we are to take seriously a "strong" form of democracy, all citizens need to deliberate at least some of the time on the decisions that affect their lives (Barber, 1984). Second, citizens participation contributes normatively to the legitimization of policy development and implementation. And third, but not least important, citizens participation can contribute to professional inquiry. Participatory forms of inquiry have the potential to provide new knowledge – in particular local knowledge – that is inaccessible to more abstract methods (Fisher, 2005).

4.2 EMPOWERING THE PUBLIC: THE JOHN DEWEY ANALYSIS OF THE PUBLIC AND ITS PROBLEMS

The three goals mentioned in the previous section – to give "meaning" to democracy, to legitimate policy development and implementation, and to feed the process with local knowledge – are at the core of the Science-Citizens-Policy Interface and of the AWARE integrated adaptive management pilot applications presented in this book.

However, the issue is hardly new. Eighty years ago John Dewey (Dewey, 1927) asked how citizens could participate in political decision making dependent on knowledge experts. Indeed, Dewey identified a paradox. As the importance of the citizen grew in the political realm – thanks to the expansion of basic rights in the nineteenth and twentieth century – the phenomenon was paralleled by the growth in power of large corporate and governmental organizations directed by managerial and technical expertise. Thus just in the period in which the political influence of the citizenry was taking shape, it was undercut by the rise of bureaucratic organization and technical expertise.

For Dewey, the answer to the challenge posed by an unprecedented level of social and technical complexity was a division of labour between the citizens and the experts. On the technical front, experts would analytically identify basic social needs and problems. On the political front, citizens could set a democratic agenda for pursuing these needs and troubles. To integrate the two processes, Dewey called for an improvement of the methods and conditions of debate, discussion, and persuasion.

Debate would require the participation of experts, but these would act in a special way: instead of rendering judgments they would analyze and interpret. If experts, acting as teachers and interpreters, could decipher the technological world for citizens and enable them to make sensible political judgments, the constitutional processes designed to advance public over selfish interests could function as originally intended in Western democracies.

4.2.1 Search for the public

The inception chapter of Dewey's masterpiece "The public and its problems" (Dewey, 1927) is devoted to the definition of the Public. This is worth a diversion, as it helps to identify the deep and robust democratic meaning of the notion of Science-Citizens-Policy interface as opposed to the more technocratic notion of Science-Policy interface. Connecting with the public very early in the policy formulation process is indeed a key requirement to ensure democracy in the policy process, and the role of the citizens is not an ancillary feature, it is at the core of the connection.

We may quote directly John Dewey: "We take then our point of departure from the objective fact that human acts have consequences upon others, that some of these consequences are perceived, and that their perception leads to subsequent effort to control action so as to secure some consequences and avoid others. Following this clew, we are led to remark that the consequences are of two kinds, those which affect the persons directly engaged in a transaction, and those which affect others beyond those immediately concerned. In this distinction we find the germ of the distinction between the private and the public. When indirect consequences are recognized and there is effort to regulate them, something having the traits of a state comes into existence. When the consequences of an action are confined, or are thought to be confined, mainly to the person directly engaged in it, the transaction is a private one. (…) The line between private and public is to be drawn on the basis of the extent and scope of the consequences of acts which are so important as to need control, whether by inhibition or by promotion. We distinguish private and public buildings, private and public schools, private paths and public highways, private assets and public funds, private persons and public officials. It is our thesis that in this distinction we find the key to the nature and office of the state. It is not without significance that etymologically "private" is defined in opposition to "official", a private person being one deprived of public position. The public consists of all who are affected by the indirect consequences of transactions to such an extent that is deemed necessary to have those consequences systematically cared for. Officials are those who look out for and take care of the interests thus affected. Since those who are indirectly affected are not directly participants in the transactions in question, it is necessary that certain persons be set apart to represent them, and see to it that their interests are conserved and protected".[1]

The problem is that in the modern and increasingly complex societies this profound connection between the officials and the members of the public – the citizens – they represent is too easily severed. Officials are indeed public agents, but agents in the sense of factors doing the business of others in securing and obviating consequences that concern them. When the public or state is involved in making social arrangements like passing laws, enforcing a contract, and so on, it still acts through concrete persons. The persons are now officers, representatives of a public and shared interest. The officials' authority is not a quality drawn from a collective impersonal will, a sort of external authorship attached to being in a power position, but the authority of recognised consequences to control the behaviour which generates and averts extensive and enduring results of weal and woe. The problem is exactly that this simple fact is too easily forgotten, because consequences of actions in complex and increasingly dynamic and interconnected societies – the latter an effect of globalisation trends already detected by Dewey in his account – are increasingly difficult to be ascertained, and the public of those indirectly and seriously affected for good or for evil is increasingly difficult to be identified.

Once more it is worth quoting Dewey about the eclipse of the public in modern societies: "Indirect, extensive, enduring and serious consequences of conjoint and interacting behaviour call a public into existence having a common interest in controlling these consequences. But the machine age has so enormously expanded, multiplied, intensified and complicated the scope of indirect consequences, and these have formed such immense and consolidated unions in action, on an impersonal rather than a community basis, that the resultant public cannot identify and distinguish itself. And this discovery is obviously an antecedent condition of any organisation on its part. Such is our thesis regarding the

[1]Dewey (1927), Chapter 1, pages 12, 15

eclipse which the public idea and interest have undergone. There are too many publics and too much of public concern for our existing resources to cope with. The problem of a democratically organised public is primarily and essentially an intellectual problem, in a degree to which the political affairs of prior ages offer no parallel".[2]

The antidote to this eclipse would be to enhance the connectivity first between scientists and citizens, and then that of the citizens with their representatives, that is concrete persons like them in a public position and responsibility. Understanding the consequences of actions is indeed a matter for science. But we need to better qualify the role of science too, as it is too easily misunderstood. The layman is led to take certain conclusions which get into circulation – for example nowadays the causes and effects of climate change – to be science. But the scientific inquirer knows that they constitute science only in relation to the methods by which they are reached. Even when true, they are not science by virtue of their correctness, but by reason of the apparatus that is employed in reaching them. This apparatus is so highly specialised that it requires more labour to acquire the ability to use and understand it than to become skilled in any other instrumentalities possessed by human beings. Science, in other words, is a highly specialized language, more difficult to learn than any natural language. As pointed out by Dewey, "it is, indeed, conceivable that sometimes methods of instruction will be devised which will enable laymen to read and hear scientific material with comprehension, even when they do not themselves use the apparatus which is science. The latter may then become for large numbers what students of languages call a passive, if not an active, vocabulary".[3]

As illustrated in the subsequent chapters of this book, the AWARE experience has shown how it is possible to connect the scientific and layman languages in an exercise of dialogue about a common concern, that in our case is the degradation of coastal waters. Citizens themselves pointed out, in their Declaration at the end of this experience, the need for more "scientific ambassadors", that is, experts in charge and really motivated to transfer scientific knowledge into knowledge brokerage processes. But besides this need to let people know and understand the complex system within which they live and operate, and become aware of the direct and indirect consequences of their actions, it is important also to acquire a critical attitude towards science and the experts' knowledge. This is particularly needed when we deal with social matters and science. Again a quote from Dewey is here particularly enlightening: "The prestige of the mathematical and physical sciences is great, and properly so. But the difference between facts which are what they are independent of human desire and endeavour and facts which are to some extent what they are because of human interest and purpose, and which alter with alteration in the latter, cannot be got rid by any methodology. The more sincerely we appeal to facts, the greater is the importance of the distinction between facts which condition human activity and facts which are conditioned by human activity."[4]

4.2.2 Connecting people

Since Dewey's time, the "eclipse of the public" question has only grown in importance. What was then a forward-looking philosophical polemic has emerged today as one of the most pressing questions of contemporary democracy theory. The progress of democracy has been disappointing. Although Western democracies exhibit high degrees of interest group involvement, levels of individual citizens participation have declined rather than expanded. Over the same period, moreover, professional experts have failed to ease the problem. Rather than adopting the role of teacher or educator, as Dewey had hoped they would, experts have largely set themselves off from the mass citizenry (Fisher, 2005).

Instead of facilitating democracy, they have mainly given shape to a more technocratic form of decision making, far more elitist than democratic. To make the matter worse, over the past decades against the increasing complexity of social and environmental problems the experts themselves became increasingly incapable to provide answers and solutions to those problems.

In recent years, this concern with both complexity and uncertainty of our problems has led influential political theorists to rethink their positions on the prospects of democracy. For instance, complexity is one of the main issues that troubles Habermas in his ongoing effort to spell out a theory of deliberative democracy. In his view, "unavoidable complexity" imposes the need for important qualifications in the elaboration of participatory democracy.

Indeed, in this age of complexity, the need for specialized expertise to formulate policies and take decisions bears directly on how much citizens can know about the choices they confront. It becomes increasingly clear that in many policy domains, politics more and more becomes a struggle between those who have expertise and those who do not. Access to technical knowledge and skill has allowed those holding the power to legitimate their political decisions. Conversely, the lack of access to such knowledge hinders the possibility of an active and meaningful involvement on the part of the large majority of the public. For instance, technical languages used by the experts provide an intimidating barrier for lay citizens seeking to express their disagreements in the language of everyday life. Speaking the language of science, as well as the jargon of particular policy communities, becomes an essential credential for

[2]Dewey (1927), Chapter IV, page 126

[3]Dewey (1927), Chapter V, page 163

[4]Dewey (1927), Chapter I, page 7

participation, which is a contradiction in terms. Not only does this directly involve the technical dimensions of policy questions, but it concerns the value trade-off and other consequences that follow from the implementation of such policies (Hill, 1992).

Bridging the Citizens-Science-Policy gap by connecting people, experts and policy makers in a shared knowledge brokerage process as that piloted by AWARE would be the recipe to counteract the eclipse of the public. There can be no public without full publicity in respect of the consequences which concern it. Whatever obstructs and restricts publicity, limits and distorts public opinion and thinking on social affairs. Without freedom of expression and systematic communication – not simply dissemination of scientific results, but truly two-way communication and common understanding of matters – not even effective methods of social inquiry can be developed. For scientific analysis and tools can be evolved and perfected only in operation; in application to observing, reporting and organizing actual subject-matter; and this application cannot occur save through free and systematic communication, of the kind that the AWARE pilot tests illustrated thereafter in this book have realized.

A last quote from Dewey helps to definitively set the problem of the public and indicate the direction to solve it: "No government by experts in which the masses do not have the chance to inform the experts as to their needs can be anything but an oligarchy managed in the interests of the few. And the enlightenment must proceed in ways which force the administrative specialists to take account of the needs. The world has suffered more from leaders and authorities than from masses. The essential need, in other words, is the improvement of the methods and conditions of debate, discussion and persuasion. That is *the* problem of the public. We have asserted that this improvement depends essentially upon freeing and perfecting the processes of inquiry and of dissemination of their conclusions. Inquiry, indeed, is a work which develops upon experts. But their expertness is not shown in framing and executing policies, but in discovering and making known the facts upon which the former depend. They are technical experts in the sense that scientific investigators and artists manifest *expertise*. It is not necessary that the many should have the knowledge and skill to carry on the needed investigations; what is required is that they have the ability to judge of the bearing of the knowledge supplied by others upon common concerns".[5] Calling in the lay citizens in the Science-Policy-Interface mechanism, the AWARE project and method build upon the shoulders of Dewey's heritage.

4.3 EMPOWERING CITIZENS' JUDGMENT: JOHN RAWLS' THEORY OF SOCIAL JUSTICE

Another important heritage that may be called in to support the AWARE experience is John Rawls' Theory of Justice and its concept of "Original Position" (Freeman, 2008; Rawls, 1999, 2005), which provides a philosophical justification to the experience and finality of the citizens juries. Indeed, *citizens juries*, if their members are appropriately selected, and the *citizens conferences*, if appropriately conducted, are the best real life approximation of the original position, the core concept of the Rawls' theory of justice as fairness.

Consideration of social justice comes in here to counter-balance – at least until the above mentioned eclipse of the public will continue to separate science and society – the special power given to the use of experts knowledge in the decision making process by the current conception of "evidence-based" policies. The point to be made here is that complexity will continue to ensure the need for professional expertise, but this only brings us to the other side of the problem; namely that the experts themselves are confronted with their own difficulties. Not only do the experts have their own professional ideological commitments, often conflicting with the public interest, but they possess no analytical wizardry capable of resolving most of the pressing societal problems. Expert judgment, at least in the field of social science, provides few uncontested solutions or answers. So, while we still need experts, expertise cannot stand alone.

Especially in social matters, normative assumptions and values are as important as technical analysis. No demonstration of efficiency can ever suffice to convince citizens to accept a social program that they don't believe to be *good*, *right*, or *fair*. When it comes to the basic normative assumptions and social understanding that underlay social and policy research itself, experts can have no privileged status. Rather than providing technical answers designed to bring political discussions to an end, the task for them would be to assist citizens in their efforts to examine their own interest and to make their own better informed decisions (Hirschhorn, 1979). Beyond merely providing analytical research and empirical data, the expert should act as a "facilitator" of public learning and empowerment.

In choices about how we want to live together – or how to solve the conflicts that arise in the struggle to do so – the experts are indeed only fellow citizens. And it is here that the case for citizens' involvement becomes apparent. More precisely, what does it mean to say that citizens (should) have a role here, empowering their judgment in the policy formulation process? Missing from Western political systems are well-developed political arrangements that provide citizens with multiple and varied participatory opportunities to deliberate basic political issues. Against this state of affairs, there is the need of a more "collaborative" or "participatory" model of expertise and policy making. Decision making procedures, in this respect, must take into consideration the authority and influence that different actors have on

[5]Dewey (1927), Chapter VI, page 208

the final choices. Should such decisions be left to the experts? What level of influence, for example, should the views of the general public carry when compared, for example, to those of scientists, administrators, elected officials, engaged community leaders, and activists? Who is more capable of judging – for instance – whether a power plant or a new regulatory program serves the interests of the public? What evidence is available? And evidence for whom (the experts alone, or the citizens as well)?

Citizens juries should be therefore formed and systematically involved in the search of answers and solutions, enabling lay citizens to participate in discussions about complex issues, including their methods of assessment (Fisher, 2005). This is also the key message the AWARE experience is bringing now to the fore. And here – to substantiate the legitimacy of citizens juries to intervene in complex policy formulation processes under given circumstances – we call in the second theoretical foundation for that experience and for the citizens conference method, which is John Rawls' conception of justice and his peculiar notion of Original Position. In the following Section 4.3.1 the rationale and features of this notion are discussed more in detail, while its relationship with the sustainability topics and the AWARE experience is discussed in Section 4.3.2.

4.3.1 Citizens' judgement and the "Original Position"

John Rawls is widely regarded as one of the most important political philosophers of the second half of the twentieth century. Rawls' theory provides a framework that explains the significance, in a society assumed to consist of free and equal persons, of political and personal liberties, of equal opportunity, and cooperative arrangements that benefit the more and the less advantaged members of society. Rawls' theory urges us to conceive a society as a fair system of cooperation over time, from one generation to the next. He says that the relationship of citizenship is a relation of citizens within a basic structure of society, a structure we enter only by birth and exit only by death. The "basic structure of society" includes the basic social institutions, such as the political constitution and framework for the legal system; the system of trials for adjudicating disputes; the norms of property, its transfer, contractual relations, and so on and finally norms that define and regulate permissible forms of the family. No society could exist without certain rules of property, contract, and transfer of goods and resources, for they make economic production, trade and consumption possible. Nor could a society long endure without some political mechanism for resolving disputes and making, revising, interpreting, and enforcing its economic and other cooperative norms; or without some form of the family, to reproduce, sustain, and nurture members of its future generations. Rawls' initial concern with justice is linked therefore to relationships between persons within an association, which are modelled using a peculiar concept of Original Position.

The Original Position (OP) is a basic concept of political philosophy. It has often been compared to the "state of nature" or the pre-political condition of humanity, which was important in the philosophies of early modern social contract theorists. According to philosophers such as Jean-Jacques Rousseau, Thomas Hobbes and John Locke, in order to understand political obligations, we should first conceive what human beings were like (or would have been like) before the creation of organized societies under government and laws, and then ask a) what reasons would have motivated people to form an organized society and b) what principles human beings in this pre-political condition would have chosen to guide their interaction in a society under an established government.

Hobbes argues that in a pre-social state of nature it would be rational for all to agree to authorize one person to exercise the absolute political power needed to enforce norms necessary for social cooperation. Contrariwise, Locke argues against absolute monarchy by contending that no existing political constitution is legitimate unless it can be contracted into without violating any natural rights or duties from a position of equal rights and equal political jurisdiction within a (relatively peaceful) state of nature.

Rawls' idea of the OP is similar to that of social contract philosophers, except that he is under no illusions that the OP was ever a reality or a "state of nature". It is a model, a *thought experiment*, that is an abstract mental device or *model* to help us understand the principles of justice that would govern the basic structure of a just and fair society. Representative (of all adult) persons in the OP are to choose principles of justice. These representatives represent every human being that belongs to the political association of free and equal persons. Due to the profoundly social nature of human relationships, Rawls sees political and economic justice as grounded in social cooperation and reciprocity.

According to Rawls, we are all "social beings" in the sense that in the absence of society and social development we have but inchoate and unrealized capacities, including our capacities for rationality, morality, even language itself. However, this does not mean that people do not have "natural" moral rights and duties outside society or in non-cooperative circumstances – Rawls clearly thinks that there are certain human rights and natural duties that apply to all human beings as such – but that they do not provide an adequate basis for ascertaining the rights and duties of justice that we owe one another as members of the same ongoing political society.

The OP perspective of justice is that, rather than representing the judgement of one person, it is conceived socially, as a general agreement by representatives of all adult members of an ongoing society. All persons – and their ideal representatives in the original position – are free and equal and their freedom consists in their possession of two moral powers: a capacity for a *sense of justice* and for a *conception of the good*. Insofar as they have these to the degree necessary to be fully cooperating members of society, they are equal. A "sense of justice" is the capacity to understand, to apply, and to act from the public conception of justice which characterizes the fair terms of cooperation. This sense expresses a willingness to act in relation to others on terms that they can also publicly endorse. A "conception of the good" includes a conception of what is valuable in human life. Normally it consists of a more or less determined scheme of final ends, that is, goals that we want to realize for their own sake, as well as attachments to other persons and loyalties to various groups and associations.

There are four fundamental features of the representatives in the Rawls' concept of the OP:

First, the representatives in the OP are *reasonable* in the sense that they wish to secure for those they represent the kind of goods that would enable them to work out their own conceptions of the good and then try to realize this good. This feature recognizes that each person has a set of interests which are of his or her own. These interests are linked to the person's moral power to form, revise, and pursue a conception of the good. Citizens are reasonable when, viewing one another as free and equal in a system of cooperation over generations, they are prepared to offer one another fair terms of social cooperation, and they agree to act on those terms, even at the cost of their own interests in particular situations, provided that others also accept those terms. For those terms to be fair, citizens offering them must reasonably think that those citizens to whom they are offered might also reasonably accept them. They must be able to do this as free and equal, and not as dominated or manipulated, or under the pressure of an inferior political or social position. Rawls call this the "criterion of reciprocity".

Second, the representatives in the OP know that a political conception of justice holds under some objective *circumstances of justice*, including physical facts about human beings, such as their rough similarity in mental and physical faculties, and vulnerability to attack, as well as conditions of moderate scarcity of resources. The latter means that the political conception of justice is applicable in social contexts where there are not enough resources to satisfy everyone's demands, but there are enough to provide all with adequate satisfaction of their basic needs; unlike conditions of extreme scarcity (e.g. famine) cooperation then seems productive and worthwhile for people. Under such circumstances, John Rawls account of justice, "justice as fairness", affirms the following principles: I) Each person has an equal right to a fully adequate scheme of equal basic liberties (a number of rights covered by the rule of law) which is compatible with a similar scheme of liberties for all, and II) Social and economic inequalities are to satisfy two conditions: a) first, they must be attached to offices and positions open to all under conditions of fair equality of opportunity; b) second, they must be to the greatest benefit of the least advantaged members of society. Principle I is named "equal liberty principle", whereas principles IIa and IIb are respectively called "equal opportunity principle" and "difference principle". The equal basic liberties have a priority, as they cannot be denied to certain social groups on the grounds that their having these liberties may enable them to block policies needed for economic efficiency and growth. For instance, a desire to increase the Gross National Product or make airlines run on efficient schedules cannot alone justify the limitation of basic liberties. Rawls' second principle means that society may undertake projects that require giving some persons more power, income, status and so on, than others – for example paying accountants and upper-level managers more than assembly-line operatives – provided that: a) access to the privileged positions is not blocked by discrimination according to irrelevant criteria (e.g. race or gender) and b) the project will make life better off for the people who are now worse off, for example by raising the living standards of everyone in the community and empowering the least advantaged persons (difference principle). What the difference principle does not permit is a change in social and economic institutions that makes life better for those who are already well off but does nothing for those who are already disadvantaged, or make their life worse.

Third, the representatives in the OP stand behind a *veil of ignorance*. That is to say, they do not know the specific details about the persons they represent: their sex, race, physical handicaps, social class, or conception of the good. They rightly assume that the persons represented have these features but they do not know what it is. This veil of ignorance deprives the parties of all the knowledge of particular facts about themselves, about one another, and even about their society and its history. The parties are not however completely ignorant of facts. They know all kinds of general facts about persons and societies, including knowledge of the relatively uncontroversial laws and generalizations that can be derived from economics, psychology, political science, and biology and other natural sciences. What they lack however is the specific knowledge of any particular facts about their own lives or other persons' lives, as well as other specific facts about their society and its population, level of wealth and resources, and so on. Rawls thinks that since the parties are required to come to an agreement on objective principles that supply universal standards of justice applying across all societies, knowledge of particular facts about any person or society is morally irrelevant and potentially prejudicial of their decision. Indeed, if the parties to Rawls' original position had knowledge of people's belief and desires, as well as knowledge of the laws, institutions and circumstances of their society, then this knowledge would influence their decisions on principles of justice. The principles agreed to would then not be sufficiently detached from the very desires, circumstances and institutions these principles are to be used to critically assess.

A fourth feature of the representatives in the original position is that they possess a great deal of *common sense* general knowledge about human psychology and sociology. They know, for instance, that humans remember the past, anticipate the future, and interact with things and people in the present. They know that people have diverse interests and talents. They are aware of the general types of situations in which humans can find themselves. They also know about the circumstances of justice – moderate scarcity and limited altruism – as well as the desirability of the primary social goods that are needed to live a good life and to develop their moral powers. They hold therefore what can be called the "everyday life" or "lay street" knowledge, something different and more general than the citizens' detailed knowledge of local circumstances which is mostly valued in participatory processes at local level.

Besides the above features, there are five formal constraints required by the concept of right, and the parties in the original position must take them into account in making their decisions. These constraints are: generality, universality in application, ordering of conflicting claims, publicity, and finality. *Generality* means that, while deciding about principles of justice, these cannot include specifics facts or features, as they should be valuable across all societies. The *ordering condition* says that a conception of justice should be able to resolve conflicting claims and order their priority. The *publicity condition* says that the parties are to assume that the principles of justice they choose will be publicly known and recognised as the basis for social cooperation among the people whose relations they regulate. Publicity of principles of justice is required to respect persons as free and equal citizens, and enable them to cooperate and live together on fair terms. Related to publicity is that principles should be *universal in application*. This implies not simply that they hold for everyone by virtue of their being moral persons. It also means that everyone can understand the principles of justice and use them in deliberations. Universality in application then imposes a limit on how complex principles of justice can be – they must be understandable to common moral sense, and not so complicated that only experts can use them in deliberations. *Finality* is related to the requirement of stability of the conception of justice and the commitment of all reasonable citizens towards it. According to Rawls, an important feature of a conception of justice is that it should generate its own support. Its principles should be such that when they are embodied in the basic structure of society people tend to acquire the corresponding sense of justice and develop a desire to act in accordance with its principles. In this case a conception of justice is stable. To be stable, principles of justice should be realizable in a feasible and enduring social world. They need to be practically possible given the limitations of the human conditions. And people should knowingly want to uphold and maintain society's just institutions not just because they benefit from them, but on the grounds of their sense of justice. A just society should be therefore be able to endure not simply as a *modus vivendi*, but for moral reasons of justice to be developed and shared by its citizens.

Summing up, whenever the features and the formal constraints are satisfied, the parties in the OP are motivated only by their own rational interests in making their decisions. Their interests are defined in terms of their each acquiring an adequate share of primary social goods – rights and liberties, powers and opportunities, income and wealth, and so on – and achieving the background social conditions enabling them to effectively pursue their conception of the good. However, since under the veil of ignorance the parties do not know their particular conceptions of the good and all other particular facts about their society, they are not in a position to engage in bargaining, and this represents a radical difference from any procedure of consultation and negotiation with stakeholders whatsoever. In effect, the representatives in the original position all have the same general information and are motivated by the same *general interests*.

In addition, it is important to note that the circumstances of the OP are not suitable for any kind of decision, but mostly for situations of ignorance and uncertainty (that scientific analysis and expert judgement would not be sufficient to reduce) about future developments which may affect the social position and well-being of the parties.

The decision at stake in the OP is not an ordinary choice. It is rather a unique and irrevocable choice where the parties decide on the basic structure of their society, or on the kind of social world they should live in and the background conditions against which they will develop and pursue their aims. It is then *sui generis*, a choice of the conditions for all future choices, and taken under ignorance and uncertainty about the future state of the world where meaningful choices will have to be made.

Under the highly exceptional circumstances of the OP – including the gravity of the choice, the fact that it is not renegotiable or repeatable, and the fact that it determines all one's future prospects – it is rational to choose conservatively to protect certain fundamental interests against loss or compromise. Once the rules of justice are decided they apply in perpetuity, and there is no opportunity to renegotiate or escape the situation, whatever a presently uncertain future will be. In this situation, it is entirely rational for the representatives in the OP to be unwilling to gamble with the basic liberties, opportunities and resources needed to pursue one's most cherished ends and commitments. The decisions that emerge from this procedure are fair and at the same time truly precautionary.

4.3.2 The AWARE citizen jury experience as a real-life approximation of the original position model

As anticipated in the previous section, citizens juries, if their members are appropriately selected, and the citizens conferences, if appropriately conducted, are the best real life approximation of the Rawls' OP model. Indeed, the real life experience of citizens conferences – and in particular of those organised in the AWARE project – fits to the OP concept in three respects.

First, the procedure applied for the selection and recruitment of the citizens jury ensures the impartiality and representativeness of its members. The actual selection of the citizens is carried out using a specific software for random selection of the participants from a pool of candidates collected through a call for applications. The call is open on Internet in principle to all the citizens in the territory of concern – the EU Member States in case of full-fledged European conferences, or the districts of a city in case of local citizens conferences – with few requirements: 1) not to hold any specific expertise of or stake in the topic of concern to avoid any conflict of interest (so scientists, stakeholders and decision makers are excluded); 2) to be available to attend a number of events (usually three workshops and a final conference scheduled over one year time, with a commitment of about 8 days of their time); 3) to be sufficiently proficient in a common language, namely English (this applies of course only for the European conferences, not those at local level). In the applications, candidate citizens are required to describe their statistical profile – age, sex, activity (employed, unemployed, retired, student), and usually one criterion related to the topic of concern (for instance, in AWARE what use the citizens more frequently do of the coastal waters, e.g. tourism, fishing, etc.). They are required also to answer two questions about: 1) how they see the topic of concern; 2) what motivates their participation. These questions aim to understand the moral standing of the participants, eliciting their own views and "conception of the good" in relation to the topic of concern. The random selection ensures the representativeness of the jury's members in relation to the population of candidates, including geographic criteria (e.g. one member of the jury for each EU Member State in the European conferences, or for each district of the city in local conferences), demographic and socio-economic distinctions, and possibly one criteria related to the specific topic of concern. An additional criterion is used to drive selection, ranking the answers of the citizens to the questions about their views of the topic and their motivations – with the help of a panel of independent evaluators – and allocating different probabilities of being selected for the citizens in the high, medium or low rank (the aim is not to exclude those of medium or low rank, but to gather a jury with a desired mix of the three categories). In this way, each selected member of the jury represents a different segment of the population of candidates. Of course, he or she does not represent in any statistically meaningful way the total population of the area of concerns, but only those of candidates. Any bias in the composition of the latter – for instance due to different access to Internet, English skills and so on – will be reflected in the composition of the jury. To enhance the representatives of the citizens jury against the total population is important therefore to disseminate the citizens participation call as widely as possible, in order to gather a population of candidates (usually at least 10 times the target number of jury's members) which is sufficiently representative of the total[6].

Second, the procedures applied throughout the whole citizens conference process (see Chapter 5) ensure equal opportunity to participate and that other OP conditions are met, namely standing behind a veil of ignorance, publicity and commitment. The members of the citizens' jury bring with themselves their common sense knowledge and moral attitudes, nothing more. They participate to the workshops and the conference as a kind of "second life" experience, far from their daily activities, cares and interests. Since the beginning of the process they are made aware of their mandate, how the process will be ruled out, and the information they will receive from the experts on the topic of concern. The whole process is carried out following a structured agenda – which may be flexible and adapted to specific needs emerging in the process, but maintains a roadmap of basic steps to be accomplished to reach the citizens conference goal. Several practical organizational requirements facilitate the process, including a suitable location, hiring a professional moderator and providing training and information as to the subject of the participation exercise in a language accessible to the citizens, arranging for lunches, dinners and coffee breaks during the meetings, and, last but not least, ensuring the individual commitment of the citizens to participate to all the meetings – continuity is a must – by signing with them letters of commitment to which are associated reasonable fees to compensate for the citizens' time use. This creates a context where the citizens act and can perceive themselves as treated on equal footing.

Third, the topic of concern should be relevant, general and serious enough to require a citizens conference process. In general, citizens participation is not always an appropriate instrument for obtaining input into policy. It is appropriate in those cases where the policy issue at stake is clearly delineated – for instance, in the AWARE project the issue was the deterioration of coastal water quality across Europe and policy was clearly framed in the EU Water Framework Directive – and can be expected to engage citizens' interests because it relates to their lives. In addition, to better resemble the circumstances of the OP, the citizens should be engaged on topics and situations in which there is a substantial ignorance and uncertainty about future developments which may affect their social position and well-being, and that of future generations. Sustainability topics, and in particular the management of common pool resources as those provided by a coastal water ecosystems (e.g. fishing grounds), are such issues. The term "Common Pool Resources" (CPR) refers to a natural or man-made resource system that provides "resource units" – fish, water, land for grazing, use of a road infrastructure, and so on – which are appropriated by multiple users (Ostrom, 1991). The resource units themselves are not subject to joint use or appropriation – they are consumed individually – but the resource system is subject to joint use, according to open access or some other socially established rule. In this situation, open access leads to the emergence of "crowding effects" and "overuse" problems, the so called "Tragedy of the Commons", as we approach an upper threshold in the number of resource units that are simultaneously consumed in the unit of time (Hardin, 1968). When the CPR is a man-made structure, such as a bridge, approaching the threshold

[6]Indeed, as pointed out in Chapter 9 of this volume, the population of candidates was sufficiently large only in the Gulf of Riga case.

leads to congestion. When the CPR is a biological resource, such as a fishery or a forest, approaching the threshold may even destroy the ecosystem. Decisions at stake in the management of CPR are therefore not ordinary choices. They are *sui generis* as the unique and irrevocable choice where the parties in the OP decide the basic structure of their society, insofar as an agreement is vital to ensure the durability of a common good. The above mentioned highly exceptional circumstances of the OP hold as well in this instance, when decisions are to be taken concerning the access and sustainable use of environmental and social resources that benefit the entire society. Moreover, there is a range of empirical solutions of "self-government" that can be found in the real world for regulating the access to a limited pool of common resources, alternative to and often more effective than both the privatization of the commons or their public management from a state authority. These solutions may engage groups of actors – for example, citizens, fishers, farmers, and so on – who are in an interdependent situation, in organizing and governing themselves, and building "bottom-up" institutions and rules of cooperation, sometime with the support of local authorities (Ostrom, 1991). To create such self-government solutions – or revise them under new threatening circumstances such as climate change – may be the subject and mandate of citizen conference processes, which eventually may evolve towards building up constitutional agreements and establish the basic rules for governing the commons. The latter may contribute to institutionalise integrated adaptive management of common resources, including in particular rules for citizens participation and empowerment, and adequate schemes for enabling a sound Science-Citizens-Policy Interface to support policy formulation, implementation and evaluation.

REFERENCES

Barber B. (1984). Strong Democracy: Participatory Politics for a New Age? University of California Press, Berkeley.

Dewey J. (1927). The Public and Its Problems. Apex Press, New York.

European Commission (2002). European Governance, A White Paper, COM(2001) 428 final (accessed 25 July 2001).

Fisher F. (2005). Citizens, Experts, and the Environment; The Politics of Local Knowledge. Duke University Press, Durham and London.

Freeman S. (2008). Original Position. Stanford Encyclopedia of Philosophy, Stanford.

Fung A., Wright E. O. and Abers R. (2003). Deepening Democracy: Institutional Innovations in Empowered Participatory Governance. Verso, London.

Giorgi L., Homeyer I. and Parsons W. (2006). Democracy in the European Union: Towards the Emergence of a Public Sphere. Routledge, London.

Habermas J. (1991). The Structural Transformation of the Public Sphere; An Inquiry into a Category of Bourgeois Society. MIT Press, Cambridge.

Habermas J. (1998). Between Facts and Norms: Contributions to a Discourse Theory of Law and Democracy. MIT Press, Cambridge.

Hardin G. (1968). The tragedy of the commons. *Science*, **162**, 1243–1248.

Held D. (1996). Models of Democracy. Stanford University Press, Stanford.

Hill S. (1992). Democratic Values and Technological Choices. Stanford University Press, Stanford.

Hirschhorn L. (1979). Alternative service and the crisis of the professions. In: Co-ops, Communes, and Collectives: Experiments in Social Change in the 1960s and 1970s, J. Case and R. C. R. Taylor (eds.). Pantheon, New York, pp. 153–193.

Jefferson Centre (2004). Citizens Jury Handbook. The Jefferson Centre, Minnesota.

Majone G. (1996). Regulating Europe. Routledge, London.

Move Together (2009). Guidelines for Local Authorities for the Organization of Citizens participation, EC FP7 Project, Brussels. Available at: www.move-together.net (accessed 10 April 2012).

Ostrom E. (1991). Governing the Commons: The Evolution of Institutions for Collective Action. Cambridge University Press, MA.

Ostrom E. (2009). A general framework for analyzing sustainability of social-ecological systems. *Science*, **325**, 419, Washington, DC.

Rawls J. (1999). A Theory of Justice. Harvard University Press, Cambridge, MA.

Rawls J. (2005). Political Liberalism. Columbia University Press, New York.

Schmitter P. (2000). How to Democratize the European Union ... and Why Bother? Rowman and Littlefield, Lanham, MD.

Chapter 5

The pilot experience with the AWARE citizens' panel

Fanny Gleize and Yves Mathieu, Missions Publiques France (F)[1]

5.1 INTRODUCTION

A group of 30 citizens, selected from three coastal areas of Europe (Gulf of Riga in Latvia and Estonia, Southern North Sea in France and Belgium, and the Goro lagoon in Italy) participated between 2010 and 2011 in the European project AWARE to give their opinion and assessment on how to improve coastal water management in Europe. Why were they given a voice on such a complex topic? Who are they? There are already experts working on it, such as scientists and water managers! What was the AWARE participatory pilot experience? What were the impacts, if any?

The AWARE approach is collaborative and solution-oriented, focusing on the dialogue between the key actors to improve the management of coastal waters. AWARE aimed to give evidence through an innovative methodology that several challenges could be met:

- Knowledge brokerage and connectivity between a variety of actors, including "lay citizens";
- Connectivity between the local and the European levels of government action and stakeholder participation (multi-governance and multi-stakeholders);
- Legitimacy of the outcomes based on the participatory process.

The AWARE citizens' panel is conceived following several methods of participatory democracy, especially developed by the Danish Board of Technology and then applied more broadly, such as consensus conferences and citizen conferences. The AWARE methodology is a hybridization of several approaches aimed at involving a small but qualified group of lay citizens in decision-making. The method used was close to a deliberative workshop, which can be described as a form of facilitated group discussion between a number of citizens selected randomly, lasting a few days. The objective of such an approach was to give them the opportunity to work in depth on an issue, where there is a mix of talks, presentations, debates leading to a collective deliberation and the production of recommendations on the issue at stake.

The AWARE group of citizens participated in several workshops at the local and European levels, where they met scientists, policy and decision makers, and stakeholders concerned with the issue of coastal water ecosystem sustainability and methods of management, at several levels of intervention. They had time to discuss and to deliberate among themselves and with these speakers, to build their ideas progressively and to elaborate their recommendations. This led to three local declarations that eventually evolved into one AWARE European Citizens' Declaration.

5.2 CREATING THE CITIZENS' PANEL

Mobilisation is the key step in a participatory process: downstream, to gather the panel; and during the process itself, to guarantee the presence of the participants.

---

[1]The authors warmly acknowledge Linda Hadfield for the English proofing of this text.

5.2.1 The recruitment of the citizens

The announcement of the AWARE citizen conference was done by means of institutional and stakeholders web-sites or other local media in the three case study areas, and via posters and flyers. 30 citizens, 10 from each of the 3 case study areas, had to be selected to compose a panel of participants sufficiently differentiated according to a number of features, including personal characteristics and motivation and interests toward the topic. The participants were recruited randomly among a list of voluntary candidates using dedicated software. Before the random selection, candidatures were analysed according to the following criteria: i) demographic (age, sex) and socio-economic characteristics; ii) different categories of water users; iii) territorial distribution; iv) motivation to commit oneself over a period of 12 months to take part in several meetings. They all had to be able to speak English (to ensure that participants from the different case studies could communicate during the European meetings). However, it is clear that this criterion excluded a large number of potential candidates (see Chapter 9 in this volume for an evaluation of this aspect).

Irina Comardicea, adelphi

The 30 selected persons were then interviewed on phone or in some cases in person, in order to explain the AWARE process. This first contact was decisive to tempt the applicants to join the adventure and to be sure they would commit themselves and attend the first meeting.

5.3 THE AWARE PARTICIPATORY PROCESS: HOW IT WORKED

A citizen conference follows 3 steps of work: 1) information; 2) investigation; 3) deliberation. Step 1 consists of providing participants with the necessary and objective information to understand the topic and what is at stake. Step 2 focuses on discussions on several topics the participants want to study more carefully, providing them with wide ranging and balanced information. Step 3 is more focused on debating the key messages and recommendations they want to convey. Even if the deliberation has started from the very beginning, this is where the participants agree on the views they share and where they dissent.

The main goal in a deliberative process is not to reach a consensus, even if it may happen, but to facilitate an exchange of views and mutual understanding. It is important to note that in the AWARE mechanism the lay citizens' deliberation process played a fundamental role for the coordination and the appreciation of a high quality research achievement. Indeed, the citizens' panel acted as:

- "integrator" of the different streams of disciplinary knowledge involved in the analysis of natural resources and ecosystems development as well as economic and social sciences for the evaluation of impacts on society. Indeed, in a deliberation setting, communication is established among scientists of different disciplines and lay citizens who hold local knowledge and multiple perspectives on the issues at stake: this requires avoiding the use of too technical and disciplinary "jargon", but instead popularising the scientific knowledge in order to be understood by both the public and the policy makers;
- "catalyser" of the connectivity established between researchers and managers, thanks to the range of social values that the public brought in to evaluate research and policy goals and outcomes/actions from the point of view of societal acceptability and usability.

The AWARE process relied on the organisation of two pan-European citizens' panel meetings – at the beginning and towards the end of the process – separated by three pilot case studies of coastal water scenario-building processes (run in parallel). The pilot case studies involved local stakeholders and policy-makers together with the AWARE team scientists and the members of the AWARE citizens panel coming from the coastal zones of the chosen case studies (local citizens). Finally, the whole process and its outcomes were disseminated by means of a final European citizens' and stakeholders conference hosted by the European Economic and Social Committee, and by following activities to spread the AWARE approach in other coastal regions of Europe.

5.3.1 From the preparation of the meeting to the first pan-European workshop in Paris

Definition of the "mandate"

> *"Over more than a year of meetings and discussions, we were asked to provide our points of view and assessment on what research tells us, what policy is doing, and how scientific knowledge, policy making and citizens' experience and values could be better connected to provide for sustainable water management choices, now and for the next generations."*[2]

This sentence, extracted from the AWARE European citizens' Declaration testifies that the whole participatory process is focused on the fulfilment of specific objectives: the assessment of a panel of European citizens on how to improve coastal water management in Europe through better connectivity between the concerned actors. Indeed, the starting point of any participatory process is to determinate the exact "mandate" or question asked to the citizens. Participants need to understand what they are in that process for and what is expected from them. It is thus important to translate the complex research issues into a clear set of questions given to the citizens' panel. The answers to the mandate have to be useful to the different partners for the scenario-building activities and to the targeted actors at the local and European levels. There was also a need to define precisely the borders of the topic submitted to the citizens' assessment. Key questions were identified and distinguished from European to local levels, for each case study. The citizens digested the issues and questions so they made sense to them and guided their work through the process. Their answer to the mandate was materialized into a European citizens' declaration; the content was determined only by the end of the 2nd pan-European citizens' meeting and, not known nor written in advance.

Preparing materials and scientific presentations on the topic

The involved citizens had to produce their advice based on the scientific evidence provided to them. Whatever the complexity of the issues at stake, this evidence had to be provided in a form that all could readily understand and use. The topic of concern was sustainable development of water ecosystems in coastal areas of Europe, devised in general and deepened by means of pilot case studies. Starting from the existing research, inception reports elaborated in the first part of the AWARE project for each case studies, and the state of the art concerning European legal framework and water governance, scientists of the AWARE initiatives worked hard "to translate" their data and analysis into simple, but not simplistic, presentations.

Scientists are also teachers; this quality was helpful to identify the preliminary information the participants might need to understand the issues at stake. The role of the facilitators involved in the AWARE consortium of partners was also a key component. Indeed, since facilitators do not hold any specific expertise regarding the issues of coastal water management and eutrophication, they could continually question the scientists and then, endorse the role of the lay citizens receiving the information. Presentations were thus prepared to fulfil the objectives of each sequence of the meeting programme. The decision of writing an English glossary including a translation of each word in French and in Italian was taken. Indeed, it was agreed that Latvian and Estonian participants would work in English during their local case study activities. Such a preparation was not an easy task for the scientists engaged in the AWARE process: the attention towards knowledge brokerage issues had to start very early indeed, from understanding and designing ways to address complex issues simple enough to be accessible for an audience of lay citizens.

Participants did not received preliminary information because it was assumed that not all of them could work on it beforehand, and this would have created inequalities in the group from the very beginning of the workshop. Also there were concerns that if they received too much information in advance, they might be intimidated by the complexity of the issues and process, and withdraw from the process before it had properly begun.

Preparing the facilitation of the citizen meetings

All citizens' debates and deliberations were followed by a professional facilitation team. Indeed, such facilitators are a third party, are neutral and constitute a warrant for the participatory process. A detailed facilitation programme had been prepared, with corresponding facilitation tools and materials. Two facilitators were responsible for the respect of timing and objectives, and their attention focused on the group dynamics and needs.

[2]Inception of the Executive Summary of the AWARE European Citizens Declaration

5.3.2 The first pan-European workshop

The 30 members of the citizens' panel gathered in Paris in spring 2010 (22–23 and 24th of April, 2010), for a first pan-European meeting. It was the first time the AWARE participants met and strong challenges had to be taken up: building up a collective that would have to last for a year at least, understanding the objectives and process of AWARE, understanding the issues at stake both at the European and local levels. Engagement and mutual confidence needed to be generated and fostered.

Objectives

The various aims of this workshop were:

- To build the group of citizens and to make them share and understand the AWARE objectives;
- To support the group in sharing their own views on coastal water management with those of the scientists, by means of specific sessions where participants were provided with the available knowledge and information on:
  - Water ecosystems functioning and management, and the issues at stake (especially eutrophication);
  - The European water policy framework and management;
  - The issue of connectivity between the different actors of coastal water management, from science, policy and society.
- To draft their first assessment of the situation in Europe.

Programme

The citizens' panel met to discuss sustainable water issues from the perspective of their everyday life experience, and the existing body of EU research which was relevant for the subsequent experiments of scenario building in the 3 case study areas. After a presentation of the AWARE initiative, the workshop started with ice-breaker activities so that the participants and the AWARE team could get to know each other. Moreover, before any expert intervention on coastal water issues and the European legal framework, the participants were asked about their own knowledge, observations, and questions regarding sea waters in the area where they live.

> *"Sometimes, we observe foam on the beach or can't see our feet through the water while walking on the beach, and we don't understand why."*[3]

It is important to note that these observations and questions the citizens raised, came from their everyday lives, discussions with family, friends and colleagues, from their own readings and information they had picked up. Their point of view as lay citizens, not experts, was as valuable as other forms of knowledge. Facilitators often reminded them to stay attached to their own feelings and perceptions when discussing with scientists or water managers, so that they felt confident enough to say what they thought and why.

Other sequences of the first workshop were mainly focused on exchanges with the AWARE project team. They aimed at producing a first understanding of water sustainability concerns and of the state of the art in research about its causes and effects. Participants also had to understand what the possible ways of acting were, or what the levers for a change of the situation were. What are coastal water ecosystems? What is the European framework and context? What is the water and coastal ecosystem situation in the French and Belgian Southern part of North Sea? What is the situation in the Sacca di Goro area? What about the situation in the Gulf of Riga?

Some work with the panellists was also about the preparation of their participation in the local integrated assessment processes (pilot case studies): the panellists were divided into sub-groups of 10 people, each to be involved into one of the pilot cases, so that they could have a deeper presentation of the issues in their area and the scenarios.

Part of the workshop was also dedicated to the concept of connectivity. Indeed, the AWARE initiative was based on the idea that sustainable coastal water management should rely upon better connectivity among all concerned actors. Participants were introduced to the complexity of the system and a mapping of all concerned actors identified around coastal water ecosystems in each case study area.

But how could the panellist digest so much information in a single meeting? What outcomes were delivered?

Moderation and results

During this workshop, the participants alternately worked in small groups and in plenary, following the prepared programme and completing assignments that allowed them to have fruitful discussions, build common grounds and thoughts, and identify the possible areas of progress and disagreement. The facilitation of the workshop was ensured by professionals (Missions Publiques, member of the consortium) with a long-lasting experience in the organization and moderation of citizens' conferences: they were not experts in coastal water management and thus assumed an external

[3]Introduction to the AWARE European Citizens Declaration

role of "translator" and "warrant" of the citizens' voice and of their progress in the knowledge brokerage process. The AWARE citizens' work was also strongly supported by the other AWARE partners, scientists as much as other task-oriented partners, who all contributed to feed the citizens' reflections and discussions.

Facilitation questions addressed to the panellists while listening to case-study presentations:

- As citizens, what are the elements that concern or interest you the most?
- As citizens, what are your doubts, or the elements that do not convince you or that you do not understand?
- Do you have any more questions?

Irina Comardicea, adelphi

No idea was lost nor judged as not interesting. As much as possible, participants were pushed to give an explanation of their ideas and recommendations. Individual and collective ideas (elaborated in small groups to facilitate the expression of every participant) were collected and finally gathered in a meeting report (elaborated after the workshop). Formal and informal moments contributed to the progressive building of a collective assessment, with a potential space for everyone's voice. By the end of the meeting, participants were asked to give their first recommendations to improve connectivity between policy, research and civil society, for sustainable coastal water systems management. A list of recommendations was made. They were finally asked to vote for ranking their recommendations in order of importance. With this meeting, a first step in the citizens' awareness and critical assessment of the topic was made. It was also clear that putting people together with scientists and managers from different areas of Europe in this pan-European meeting facilitated the brokerage of experiences and ultimately triggered the transfer of knowledge between all types of persons present.

5.3.3 The local processes

During the autumn 2010, the AWARE citizens met at local level, divided into 3 local groups they joined according to the pilot coastal areas to which they belong. There, they went into a deeper analysis and assessment of the local coastal water management issues, by discussing with local stakeholders. Goals of the case studies were indeed to openly discuss various interests and problems identified and produce scenarios for the future of water use and quality in the coastal zone. This approach promoted a fruitful exchange of information among scientists, managers and the public, and the true integration of diverse stakeholders' knowledge and social values into the process.

A 2-day local workshop took place in parallel in the 3 concerned areas, gathering each 10 panellists, the AWARE scientists and key local stakeholders and managers, with the aim of discussing scientific findings, connectivity issues and the development of scenarios. The scenarios concerned the deterioration of water in coastal areas and policies to avoid or mitigate these phenomena.

The topics at stake and the list of invited local stakeholders were debated among case study partners, also according to the answers to an online survey on local coastal water ecosystem sent earlier in summer 2010. The online survey was useful to identify different stakeholders' opinions and invite some of the respondents at the local workshops, to meet and discuss their views with the local panellists.

The workshop was also facilitated by professional moderators involved in the AWARE consortium, to structure and support the discussions (by means of work in small groups at given times, systematic production of ideas/remarks/recommendations after discussions/speech, clear speeches given at scheduled times by invited experts and stakeholder representatives, facilitated round-tables, etc.).

Participants wrote a local Citizens' declaration addressed to local decision and policy-makers, stakeholders and scientists, and presented it during a local conference (all 3 Declarations are available on the AWARE website www.aware-eu.net). This local conference occurred a few weeks after the workshop (with the exception of the Gulf of Riga case, where the conference took place the following Monday after the Friday-Saturday workshop), to provide sufficient time to incorporate public comment gained during the previous local workshop.

Such participatory scenario-building activities held in parallel in the 3 case-study areas revealed that coastal water management suffers from a lack of sufficient spaces for dialogue between all concerned actors. Citizens involved, by their reflections and questionings, actually played a role of "mediator" between research and policy development, in each specific case study circumstances.

Focus on the Southern North Sea case study citizens' process[4]

The local workshop

On the 8th-9th of October 2010, the French and Belgian citizens of the AWARE project took part in a workshop hosted by the Université Libre de Bruxelles (ULB). Together with scientists and local policy actors, these ten citizens discussed the environmental situation of the southern North Sea coastal marine waters. During two full and intensive days, the citizens had the opportunity to discuss their ideas with local policy-makers, key institutional representatives, as well as NGO representatives and local stakeholders from tourism, farming, and fisheries, all from France and Belgium. Four sequences of discussion were organised as round-tables of speakers about the following specific topics:

Round-table 1: Impact of eutrophication on fisheries and tourism in the North Sea. Comparison with other nuisances. Testimonials of different stakeholders.
Round-table 2: Governance: political framework, legislation and directives related to eutrophication.
Round-table 3: Point sources of nutrients: origin, reduction, management of waste water.
Round-table 4: Diffuse sources of nutrients: origin, reduction, rehabilitation of wetlands, future agriculture.

The 1st round-table aimed at understanding what the environmental, social and economic impacts of eutrophication were, by listening to concrete testimonies of actors directly concerned. The 2nd round-table was rather focused on the existing connectivity or, at least, aspects of governance at local level. The 3rd and 4th round-tables were solution-oriented: the first one addressed solutions on point sources of nutrients (mostly action on urban waste-water treatment); the second one mainly addressed actions related to land management and agriculture.

As in the first plenary workshop in Paris, participants discussed with the speakers and the scientists, alternately in plenary group or in small groups; they also always worked on the ideas, questions and recommendations they had all along the workshop.

The Declaration and scenarios

During the workshop, the ten participants exchanged their views and developed opinions on solutions to improve the management of the water ecosystem in their coastal area. By the end of the workshop, the citizens called for a better coherence and an increasing flexibility for action in the efforts to fight against eutrophication, as well as the involvement of all actors of the society (including citizens). They also queried whether investments to address point sources pollution were not coming close to saturation level (both in terms of cost and processing techniques). Thus, they put as a priority "to act in allocating resources to fight against diffuse sources including agriculture." However, they called for a better recognition of the role of farmers instead of pointing at them as responsible for all the pollution in waters. Indeed, they thought to have a direct responsibility as citizens, in their everyday life and in their consumption habits, towards the whole ecosystem and its health.

The Citizens' Declaration for a sustainable management of coastal water ecosystems in the Southern North Sea and Seine, Somme and Scheldt river basins focuses on three main themes:

- Transparency/Consistency/Dynamics;
- Information/Openness/Education;
- Diffuse source/Consumption.

[4]Details concerning the other case studies processes are provided, respectively, in Chapter 6 for the Gulf of Riga case and in Chapter 8 for the Sacca di Goro case. Chapter 7 illustrates the scientific content and outcomes of the Southern North Sea case study.

During the workshop, the ten participants had also the opportunity to discuss with the scientific team involved in the project and provide them with some recommendations for their environmental models and scenarios of reduction of nutrients at the coastal zone. The scientific teams from the Université Libre de Bruxelles and Université Pierre et Marie Curie integrated that input in their work. They presented their results publically at the AWARE Local Conference in Dunkerque on 7 January 2011 (see Chapter 7).

The local Conference

The conference welcomed about a hundred participants, mostly key stakeholders in the water and related (i.e. agriculture) sectors. Such a success was reached thanks to the partnership built with the Urban Community of Dunkerque Grand Littoral, which supported the AWARE initiative both providing original content (the area of Dunkerque often faces eutrophication consequences on the beaches) and helping to organise the local conference and reach the local stakeholders. This partnership was essential to perform a wide but targeted mobilisation, but also to reach key decision-makers with the citizen and scientists messages.

The citizens presented their Declaration and discussed it with the participants, who seemed to be positively surprised by the quality of the reflections and the questions raised. The scenarios explored by the AWARE scientific team were also presented, after they were elaborated based on the Brussels workshop discussions with citizens and stakeholders. Their modelled scenarios explored some realistic and more visionary nutrient reduction measures for reaching a better quality of the continental and marine waters. They especially developed 3 scenarios:

- Improving wastewater treatment: if the European Directive dealing with the improvement of waste water treatment plants (91/271/CEE) is fully implemented;
- Reducing nutrient losses from agriculture (action on diffuse sources of nutrients): if drastic changes such as organic farming and diet change (50% decrease of animal protein in human diet) are made;
- Managing the water landscape by implementing artificial wetlands (ponds or reservoirs have the capacity to retain and eliminate nitrate).

There were active interactions with the audience focusing on scenario outcomes as well as on the participatory scenario-building process applied to elaborate them. The feasibility and outcomes of different measures to improve water quality, and the importance of increasing the connectivity amongst water actors with initiatives that foster a real dialogue were at the core of discussions. The gap between politics and stakeholders, and scientists and citizens, must be further reduced. The apparent complexity can be simplified if the knowledge does not remain frozen, but on the contrary, shared and mobilized. The communication of scientific knowledge to the broad public was identified as one of the key and, at the same time, weakest point, to improve water management. The conference was, thus, a common learning process for all the actors involved where they shared knowledge, opinions and tools.

5.3.4. The second workshop, a year after

In spring 2011 – exactly a year after the 1st European workshop, the AWARE group (30 citizens) gathered again, to meet a second time in Paris (28-29-30 of April 2011). The challenges of this meeting were mainly to make the link between the local and the European levels, by sharing the outputs of each of the three case studies, and to formalize their recommendations in a European citizens' declaration.

Objectives

The detailed objectives of this 2nd workshop were:

- To share the 3 local processes and results, and extract common conclusions;
- To complete the training and discussions with scientists, stakeholders and policy makers on sustainable coastal water management in Europe;
- To finally produce their assessment and formulate their recommendations on new modalities of linking research results to policy-making, based on the case studies' outcomes and targeted at decision and policy makers at the European level, stakeholders, scientists, civil society organizations and the public at large.

Programme

After having worked in subgroups of ten to produce recommendations regarding sustainable coastal water management at the local level, the thirty citizens came together in Paris to draft a common declaration addressed to European policy makers. Citizens, as well as scientists involved in the AWARE initiative, were very motivated to share their feelings and ideas about the participatory process they had experienced at the local level, and their recommendations for the European decision-making level.

Irina Comardicea, adelphi

Local citizens' declarations were sent a few days prior to the meeting so that all the citizens could know what were the reflections and recommendations from each case study, and think about what surprised them in that process, what were the similarities, what were the differences. After sharing their thoughts, a session was organized with European stakeholders for an update on the European water framework, as well as a state of the art of the science and policy interface (see Chapter 2 in this volume). Similar interaction methods were used to capitalize on what they retain, what they question and what they recommend regarding the speeches they listened to. As agriculture was a common topic discussed in the three case studies, a representative of the agricultural sector was invited to discuss with the citizens on possible solutions which could be implemented to respond to the problems of water quality. Finally, as information and communication issues were recurring in all three citizens' declarations, a special session was dedicated to a critical review of communication tools and experiences, with the aim of fostering the participants' attention and focus on the lack of sound communication among all concerned actors, and to allow them to provide recommendations for the European level. Thus, the first day of the 2nd pan-European workshop was dedicated to thematic sessions of discussions, while the second day was fully focused on the deepening of ideas, recommendations and justifications to them. Participants were not pushed to formally write down the Declaration – this was done afterwards – but rather to agree collectively on what they agreed or disagreed. The structure and content of the European Citizen Declaration finally emerged from this second intensive workshop day.

Moderation and results

To produce the final declaration, participants agreed in plenary group upon the main structure of the text and split into small mixed groups composed of citizens from each of the 3 case study areas, to work on one assigned part of the text. After a while, each group presented to the others the content of its work. It was commented and amended in plenary so that the declaration could reflect what the whole group thinks of the issue, and not only the 4 or 5 people concerned in each small sub-group. When disagreements were encountered, facilitators were not pushing to a consensus but rather encouraging them to develop the argument, give examples, and so on. As already said, there could be an agreement on disagreements, where different opinions could be expressed but explicitly justified. It is clear that when policy-makers receive a citizen declaration, they don't know what to do with judgments and advice with no explanations behind them. The impact may be stronger with clear examples and reference to observations and thoughts they have, as citizens, or referring to what they experienced in the AWARE process, than with general judgments or sentences retained directly from a speaker or another. The task was difficult but the citizens were highly motivated by the wish to be taken seriously into consideration in Brussels, at the European public conference, but also by any future potential reader of their declaration. Scientists of the AWARE team were also very helpful, in particular to advise the citizens when needed to formulate in the most correct way what they retained from the eutrophication process and models for solutions, without changing however their ideas or writing in their place.

A few weeks after the workshop, the European AWARE Citizens' Declaration was finalized, gathering their assessment on how to achieve better sustainable coastal water management in Europe, and their recommendations to improve connectivity between all concerned actors. The citizens also produced an Executive Summary in order to reach their audience at the European Economic and Social Committee on the 9th of June 2011, and call all concerned actors and decision-makers to take their work into account (and act!).

5.3.5 The European conference

"Linking research to policy in the water sector: connecting people for a sustainable coastal water management in Europe" was the title of the one day conference organised with the support of the European Economic and Social Committee. By sharing the AWARE experience, the conference had many challenges to take up, as its title suggests, calling for an improvement in the science-policy connectivity in the water sector and for more public participation in the context of EU policies. The EESC played an active role in the conference programme and mobilisation of the audience, being a European Institution mainly concerned with the involvement of civil society on EU policy issues in different sectors, including natural resource exploitation. The most relevant policy makers and stakeholders (represented by their Brussels offices), but also representatives of associations and academia were especially targeted. The AWARE citizens, scientists and other AWARE partners presented their full recommendations to an audience of about 70 people. Some stakeholders or policy-makers involved in the AWARE initiative also testified both on coastal waters problems and issues of connectivity in the water sector. Lively discussions took place, with the audience sharing opinions on water management and public participation. Perspectives on the AWARE initiative were given by representatives of the EESC, of the Committee of Regions and finally by the European Commission, Directorate General of Research and Innovation.

Irina Comardicea, adelphi

This public conference was an essential step in the AWARE process, as it was the time when decision-makers and stakeholders were officially invited to give their reactions to the citizens' reflections and recommendations. The invited decision makers were asked to give not necessarily a positive (and potentially demagogic) answer to every recommendation, but to offer their insights about the usefulness of including the citizens knowledge, as elicited in the AWARE process, as a complementary perspective in the water policy formulation process, at EU and local level, and to explain what can be done or not, and why. This dialogue between the public on one hand and the decision-makers on the other hand was the climax of the AWARE participatory experience.

5.4 THE CITIZENS' MESSAGES
We want to be part of water governance

"We – the citizens- are too often out of the system."[5]

The AWARE citizens called strongly for recognition of the role of citizens in the decision-making process, not only during elections or referendums. They observed that some efforts are made to open the debate to the public, as is the case in the Water Framework directive. But they noted that, in fact, areas of debates are not done to involve, inform, listen to lay citizens, nor take their voice into account. Usually, public participation is recognized through the action of the organised civil society, like associations. However, citizens do not always recognise themselves in the actions or messages of such associations. What the AWARE citizens experienced during a year has shown that lay citizens – if they are informed in a balanced and pedagogic way – can produce strong arguments and valuable ideas. Their expertise is as valuable as that of scientists, technicians and politicians, and is fully complementary, adding a different perspective in the knowledge brokerage process. Finally, what they question here is the sharing of power and the need to find together a common interest so they all sit around the same table.

[5]Inception of the European citizens' declaration

The citizens' key recommendations: executive summary of their European citizen Declaration

"We are a group of 30 citizens coming from five European countries who volunteered and were selected randomly to participate to the AWARE European project – a project that connects people for better European coastal water management (www.aware-eu.net). During the project we discussed three very different case studies (the Sacca di Goro lagoon in Italy, the Gulf of Riga in Estonia and Latvia, and the Southern North Sea and the Seine, Somme and Scheldt river basins in France and Belgium). Over more than a year of meetings and discussions, we were asked to provide our points of view and assessment on what research tells us, what policy is doing, and how scientific knowledge, policy making and citizens' experience and values could be better connected to provide for sustainable water management choices, now and for the next generations.

How can we achieve a better and more sustainable coastal water management in Europe? Here is a summary of our main conclusions:

ANOTHER APPROACH TO UNDERSTAND COMPLEXITY

Throughout AWARE we focused on the challenge of eutrophication, but it is clear to us that issues of water quality cannot be restricted to that one challenge alone. We observe that *eutrophication is in fact only one point of entry into broader ecosystem health issues*, which include pollutants coming from both the land and the sea. We underline that only a *holistic, or multidimensional, approach* can help all involved actors (citizens, stakeholders, experts from different disciplines, decision makers) understand the complex issues surrounding water quality. Only such an approach can receive acceptance from relevant actors, and lead to sustainable and long-term solutions.

AN UPDATE OF THE WATER FRAMEWORK DIRECTIVE (WFD)

We realised that there has been little room given to citizens so far in the implementation of the WFD. We believe we are legitimate actors that should be part of the decision making process defining what "good" environmental status is, and given the chance to share our opinions with scientists and key policy-makers. *We – the citizens – are too often out of the system.* We have heard evidence that the European Water Framework Directive's goals may not be reached in many areas of Europe. We are concerned about the quality of our coastal waters and by the lack of consistency we observe in water quality data, in the monitoring of water quality, and in the enforcement power of the European Union. We call for an updated Water Framework Directive, which will allow for better continuous and participatory improvement and strict control of the implementation of the Directive, as well as for *more consistency in the measurement of data, and between different EU policies*. For example, could DG Environment of the European Commission, as working on a transversal topic, be playing a bigger role in making bridges between the WFD and the Common Agricultural Policy, or between the WFD and the Marine Strategy?

TIMELY, LOCAL, AND RELEVANT SOLUTIONS ON THE GROUND

We know that we are currently still seeing the effects on coastal waters of several decades of human practice and we know that the ecosystem has a strong capacity of inertia: change will come slowly. This is why the time for talking and for standing still has passed, and *the time for actions has come*. We are aware that no single solution will be the best per se. We strongly recommend that all options be evaluated against *environmental and socioeconomic criteria* – especially those relevant at the local level.

It is clear to us that challenges are different for every region and we therefore encourage *solutions that make sense and are most effective at the local level.* One region may benefit from increasing the number of waste water treatment plants, but supporting a transition to organic farming may be the best for another area.

BETTER CONNECTIVITY BETWEEN RESEARCH, POLICY-MAKING AND CIVIL SOCIETY

Throughout the AWARE process, we observed and experienced important gaps between all concerned actors. According to us, these gaps jeopardize the achievement of sustainable water management in Europe. It has become clear that expert information both from researchers and from practitioners are necessary to take better, sustainable decisions on water quality. Scientific experience and consultations are crucial. But so is the information from citizens, farmers, fishermen, and other involved actors. *Scientific ambassadors* in fact, would be a great way to "translate" critical information for citizens, business representatives, and politicians alike – as it happened with the information we benefited from in the AWARE process. We also push for better connectivity between all concerned actors at the *water basin scale*, across borders, *including non EU member states when necessary.*

BETTER INFORMATION TO KEY ACTORS

Citizens are not the only ones who should benefit from better and clearer information on water quality issues. Other stakeholders also need to receive full information in order to help them make better choices. *Dialogue with farmers, fishers, tourist organisations* and other stakeholders should be strengthened and reinforced. We as citizens can approach farmer lobbies for instance, but scientists also need to enter into a *dialogue with the citizens* about coastal water health and upstream challenges. A special focus should also be placed on *children's education* because they also need to act now, and in the future. Because *policy makers are the ones who can truly make* change happen, we encourage them to listen to citizens and scientists to have the most enlightened and appropriate decisions.

What are we ready to do as citizens?

We are ready to be part of a change. We recognise our share of responsibility for the impact on the environment and therefore we are willing to act – as consumers, as part of the decision making process, alone or collectively.

Although politicians, business and scientists more often communicate with each other through existing channels (like in the Science-policy interface) we found that *citizens are often left out, or information about water quality issues does not reach us.* Keeping us, citizens, in the dark prevents us from making full use of our ability to contribute to the decision-making process. We remind that with the right type of expert information we were able to say precisely what we are ready and not ready to do, and explain why we hold those opinions. *Including our opinions would result in more sustainable water management decisions.*

"We as citizens are willing to take the first steps and hope other parties will follow. Join us in this change"

Declaration presented on the 9th of June 2011 at the EESC.

Irina Comardicea, adelphi

We are ready to act

> *"We recognise our share of responsibility for the impact on the environment and therefore we are willing to act – as consumers, as part of the decision making process, alone or collectively."*[6]

The AWARE citizens also consider that they are not only spectators of environmental problems but actors. Citizens are not only end-users as they are too often presented in the water governance. They are not only paying their water bill. They are at every end of the system; they can act from the very beginning of the ecosystem cycle, through their consumption behaviour for example that will have consequences on agriculture or urban water treatment. As any concerned actor in the ecosystem cycle, they endorse responsibility and willingness to undertake personal actions aimed to improve the situation.

We do care about the environment, the quality of coastal waters is just one entry

> *"It is clear to us that issues of water quality cannot be restricted to the challenge of eutrophication alone. We observe that eutrophication is in fact only one point of entry into broader ecosystem health issues, which include pollutants coming from both the land and the sea."*[7]

The AWARE citizens called for a global or holistic look at water quality problems; the issues are much larger than the one of eutrophication. They do care about environmental problems but they are also concerned by social and economic impacts. Issues have to be taken into consideration both downstream, in the coastal areas, and upstream, in the river basins. For them, without clear scientific and socio-economic information on the environment and on policy options, no sustainable decisions can be taken by the involved actors – citizens, politicians, scientists, fishermen, tourist organisations, farmers, and so on. Related to a global approach, the corresponding connectivity amongst actors should happen; they were indeed shocked to observe important gaps between all concerned actors that jeopardize the achievement of sustainable water management in Europe. *"We push for better connectivity between all concerned actors at the water basin scale, across borders, including non EU member states when necessary."*[8]

---

[6]Ibid.

[7]Ibid.

[8]Ibid.

5.5 WHAT DOES THE PARTICIPATORY PROCESS PRODUCE?

5.5.1 The recognition of several expertises and knowledge brokerage

The AWARE participatory process illustrated in the previous sections has demonstrated that lay citizens, being provided with balanced information, can discuss and formulate relevant recommendations on complex issues such as water quality or eutrophication – a word they didn't even know before being part of this project. They formulated recommendations addressing high level representatives, in order to influence decision-making at EU and local levels. There is indeed a huge lack of space for dialogue in our societies, as was underlined by the citizens.

The AWARE participatory process not only gave voice to citizens, but also to scientists in knowledge brokerage activities. Step by step, the AWARE scientific team became more and more open to the reflection and interactions with the citizens. They worked hard on their presentations and models to make the material as pedagogical as possible. They opened their knowledge to disciplines, such as socio-economic impact assessment, they were not accustomed to, in the perspective of enriching their research and the scenarios of improvement of the situation in their case study. It was clear that the work with easily comprehensible scientific presentations and the participatory building of scenarios proved to be useful communication tools. The AWARE scientists could also enter into a dialogue that was not so common to them, with stakeholders and decision-makers; they could make their voice heard as well as understanding the issues of concern for these actors.

Besides the citizens' panel, many categories of stakeholders were engaged in the process, at local level and in the EU conference, including authorities involved in river basins and/or coastal water management, interest groups (e.g. farmer unions), business representatives (e.g. tourism, farming, fishing, etc.) and Civil Society Organisations (CSOs) active in the water sector, and of course the water scientists as well. As it concerns the water experts and practitioners, the AWARE process has let them better understand lay people motivations for behavioural change and action. Moreover, parallel and complementary activities in the AWARE project – like the online surveys and interviews – which engaged representatives from all the above mentioned categories of actors, have shown opportunities to foster dialogue but also persistent cultural barriers which make dialogue and especially the inclusion of the citizens in the science and policy interface mechanism, still difficult to implement. This would require indeed a significant shift in the current mentality of most scientific experts and decision makers alike.

5.5.2 Added-value to local and European policies

"More legitimacy through process"[9]

The AWARE initiative invented a rigorous process to develop a collective point of view of European citizens on sustainable coastal waters. The process started with a group of citizens chosen at random, and step by step, through dialogue with scientists and a variety of stakeholders, led to the elaboration of reflections and recommendations valuable both at local level (in three distinct areas) and at European level. AWARE proved that a same group of citizens could work both at local and European levels, and that these two scales of reflections were not only encouraged but necessary. Local processes were indeed essential to elaborate consistent recommendations at the European level. Otherwise, the risk was to maintain a too general and declarative speech mode of thinking, full of good advice and judgments but far from being concrete. European policies are so complex, but the citizens need to see how European Directives are translated at local level and which impacts deliver at the scale of people's everyday life.

The AWARE outcomes have gained legitimacy through the participatory process and the dialogue organised with a wide range of actors; the legitimacy comes also from the articulation between the local and the European scales, which gave more concreteness to the whole process. Finally, the knowledge brokerage process has demonstrated that citizens often hold – from their everyday life perspective – a less specialist but more global and systemic vision of the issues at stake, while public officials often elaborate policies from sectoral perspectives, which often do not take into account side effects and unintended consequences that may appear clear instead to affected lay citizens.

Summing up, AWARE may represent a solution to bridge the gap between citizens, science and policy, but only if an impact is really obtained on policy-making (see in this respect the weaknesses highlighted by the AWARE evaluation team, reported in Chapter 9). However, the main goal of the whole AWARE process is not to deliver recommendations to be applied exactly as they are, but rather to establish a framework for systematic discussion with policy-makers, at EU and local levels, about their constraints and opportunities for acting. The politicians should benefit by listening to the citizens, and acquiring their factual knowledge of the subject matters. Citizens, on their side, may better hold politicians accountable and demand long term commitments. They especially call to be heard and recognized. Also, there is now a concrete hope that an increased use and dissemination of sustainable water research results will facilitate water policy formulation and planning in the case study areas of the AWARE project, and elsewhere in Europe by means of emulation.

---

[9]Derenne, B. (2006–2007) *"What roles for rural areas in tomorrow's Europe?" Regional and European perspectives from the European Citizens' Panel Initiative*, Complete results. Final publication may be downloaded at www.citizenspanel.eu

Chapter 6

The future of the Gulf of Riga: pollution, water quality and fish production

Andreas C. Bryhn[1], Kristina Veidemane[2], Per Stålnacke[3] and Udaya Sekhar Nagothu[3]

[1]Swedish University of Agricultural Sciences (SLU), Department of Aquatic Resources, Skolgatan 6, 74242 Öregrund, Sweden and Uppsala University, Department of Earth Sciences, Villavägen 16, 75236 Uppsala, Sweden
[2]Baltic Environmental Forum, Antonijas 3–8, Riga, Latvia
[3]Bioforsk-Norwegian Institute for Agricultural and Environmental Research, FA Dahls vei 20, 1432 Ås, Norway

6.1 INTRODUCTION

In this chapter, the Gulf of Riga (GoR), one of the case studies chosen in the AWARE project is described in detail. The GoR waters are shared by five countries and thus being a transboundary water body, offers a major challenge to water resources management in the region. The regional differences in economy and political set up also add to the management challenges and the implementation of holistic mitigation measures in the GoR. These are some of the main reasons for lack of a good stakeholder engagement process involving different countries and also developing water management policies based on sound scientific inputs. In the AWARE project, attempts were made to look at the current gaps in the stakeholder involvement and the citizens-science-policy interface and how involvement and interfaces can be improved. This chapter provides a brief introduction of the geography, followed by a description of the eutrophication and fish stock fluctuations in the Gulf, which has been identified as one of the main problems in the region. The modelling approach used in the AWARE project will subsequently be described and finally the experience on the participatory citizens-science-policy process is discussed.

6.2 CASE STUDY AREA

The Gulf of Riga (58°N, 23°30′E; Figure 6.1) is a shallow sub-basin of the Baltic Sea. It has a surface area of 16 700 km^2, a mean depth of 24.5 m and a maximum depth of 56 m (Håkanson & Bryhn, 2008a). The water of the Gulf is brackish as the salinity in most parts ranges from 5.5 to 6.0 psu (Kotta *et al.* 2008). The Gulf of Riga is bordered by the Estonian and Latvian mainlands, the two islands of Muhu and Saaremaa, the Irbe Sound and the Suur Strait. The main anthropogenic pollutants to the Gulf are nutrients (Kotta *et al.* 2008) from domestic sewage, agriculture and industries. Most of the nutrient pollution loading into the Gulf of Riga can be attributed to activities in the drainage basin and inputs from the adjacent Baltic Proper. The Gulf has a total drainage area of 138 000 km^2 (Figure 6.2), including parts of Estonia (7 000 km^2), Latvia (59 000 km^2), Lithuania (10 000 km^2), Belarus (38 000 km^2) and of Russia (24 000 km^2).

More than 4.6 million people live in the drainage basin of the Gulf. The population is quite uniformly distributed among and within the river basins with the exception of a few large cities. Between 1950 and 2005, the population increase rate along the coasts of Estonia and Latvia exceeded the total population increase rate in these countries by more than 50% and the population grew substantially in most parts of the whole north-eastern Baltic Sea region during this time (Looring, 2009). Forty percent of the drainage area (mainly lowland) is used for agricultural production. With the increase in population, the pressure on land and water resources in the GoR has also increased significantly over the years. As a result of heavy human pressure, the GoR has been subjected to pollution from agricultural and urban sources leading to eutrophication and other problems.

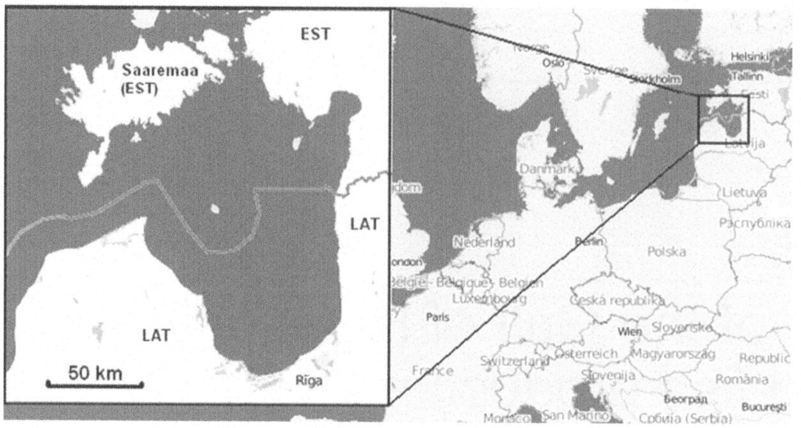

Figure 6.1 The Gulf of Riga and its location in Europe. EST: Estonia, LAT: Latvia.

Figure 6.2 The Gulf of Riga drainage basin with the 5 major rivers and river mouth monitoring sites (triangles). *Source*: Stålnacke *et al.* (1999).

6.2.1 Eutrophication and fish stocks in the Gulf of Riga

Eutrophication is often regarded as the major environmental problem of the Baltic Sea (Enell & Fejes, 1995; Hassler, 2003). Anthropogenic nutrient enrichment has been prevalent for many decades and has been subject to research and restoration efforts since the 1960s. Nutrient loadings from the catchment and the atmosphere today are lower than the levels recorded in the 1980s although the water quality has not improved since then (Neumann & Schernewski, 2008; Savchuk & Wulff, 2009; Bryhn, 2010). Eutrophication is commonly manifested as high nutrient concentrations, low Secchi depth (water transparency) and intensive algal blooms.

The Gulf of Riga is one of the most nutrient polluted sub-basins of the Baltic Sea (Håkanson & Bryhn, 2008a; Kotta *et al.* 2008). The average Secchi depth in the Gulf has decreased from about 5 m in the early 1960s to 3 m during the first years of the 2000s (Fleming-Lehtinen & Kaartokallio, 2009). Phosphorus loadings may have doubled from 1950 to 2000 during the population increase mentioned above (Looring, 2009).

Rivers transport account for most of the total pollution load from the drainage basin to the Gulf of Riga. More precisely, approximately 113 000 tonnes of total nitrogen (79% of the total load) and 2100 tonnes of total phosphorus (68% of the total load) were transported by the rivers to the Gulf in an average year during the period 1977–1995 (Laznik *et al.* 1999). The total phosphorus (TP) flux from the Baltic Proper to the Gulf of Riga is about 4,300 tonnes per year, according to two separate estimates (Savchuk, 2005; Håkanson & Bryhn, 2008a), and is thus larger than the combined loading from the drainage basin. Concentrations of the nutrient phosphorus (P) regulate the long-term growth level of phytoplankton in the Gulf and an increase in phytoplankton abundance (including cyanobacteria) has been noted since the 1970s (Kotta *et al.* 2008). The slightly increasing trend in total phosphorus (TP) concentrations is illustrated in Figure 6.3 regarding the years 1986–2007. In the late 1980s, concentrations varied around 27 μg/l, while concentrations varied around 29 μg/l in the 1990s and were near 31 μg/l during 2000–2007 (Bryhn, 2009).

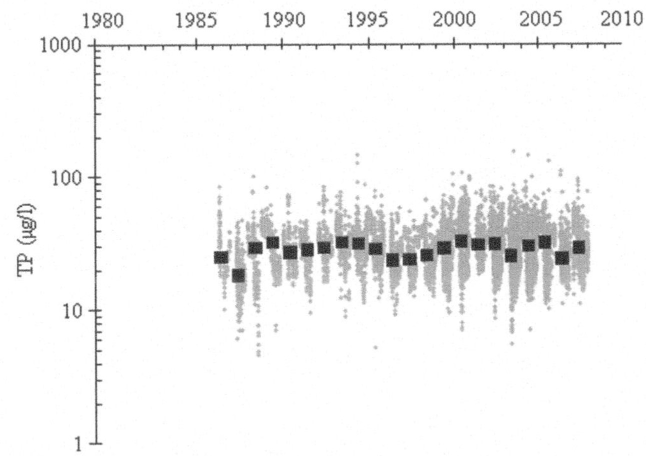

Figure 6.3 Total phosphorus (TP) concentrations in surface waters of the Gulf of Riga. Black squares denote yearly medians. Number of data: 7441. Trend slope 1986–2007 was +0.097 µg/(l · year), $r^2 = 0.002$, $p < 0.001$. From Bryhn (2010).

Nutrient concentrations and phytoplankton biomass are also connected to fish productivity. The higher the primary production the higher the fish yield (Houde & Rutherford, 1993; Chassot *et al.* 2007; Bryhn, 2012). This general relationship is displayed in Figure 6.4, which illustrates a central fact related to measures to reduce eutrophication. The fish productivity of water bodies is connected to primary production by many intermediate trophic links (see Håkanson & Boulion, 2002; Chassot *et al.* 2007). Although some effects from eutrophication (oxygen depletion, toxic phytoplankton blooms, shrinking macrophyte colonies) can disturb fish activity, the dominating net effect is that the amount of food for fish increases and thereby fish production (Bryhn, 2012). The relationship in Figure 6.4 suggests that if eutrophication is greatly reduced by costly remedial measures, it is likely that the fish production will also be lowered which would affect fisheries and livelihoods of people dependant on fishing in the Gulf. Patterns similar to those depicted in Figure 6.4 have been documented by Nixon (1982), Iverson (1990), Ware and Thomson (2005), Chassot *et al.* (2007) and Jennings and Brander (2010). The Baltic herring (*Clupea harengus membras)* is the most abundant among more than 50 other species in the Gulf and herring also accounts for about 90% of the commercial catch value. The herring stock is quite stationary in the Gulf (Kotta *et al.* 2008) and Figure 6.5 shows the annual variations of landings and spawning stock biomass of herring in the Gulf of Riga 1977–2007.

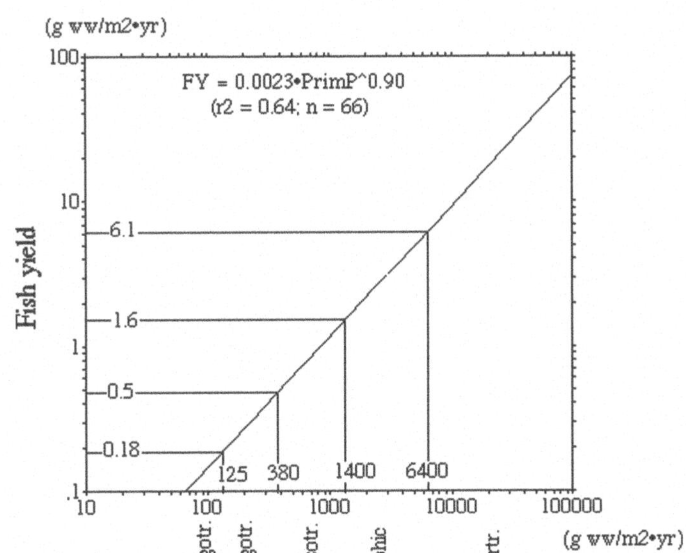

Figure 6.4 A cross-systems correlation between primary production (trophic state) and fish yield. From Håkanson and Boulion (2002).

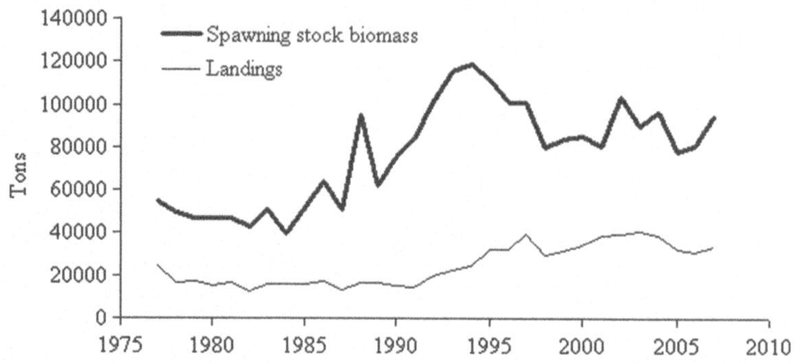

Figure 6.5 Spawning stock biomass and landings of herring in the Gulf of Riga, 1977–2007. Data from ICES (2008).

One can note a strong increase in spawning stock biomass of herring between the mid-1980s and the early 1990s in Figure 6.5, a change which concurred with increasing nutrient concentrations and primary production in this area (Kotta *et al.* 2008; Bryhn, 2009). A complementary and perhaps even more important explanation to this increase may be the decline in abundance of a temporarily important predator species in the Gulf, the Baltic cod (*Gadus morhua callarias*). Large, piscivorous cod were much more prevalent in 1977–1985 than during subsequent years (Ojaveer *et al.* 1999; Kotta *et al.* 2008). Since the early 1990s, herring landings have increased while the spawning stock biomass appears to have stabilised (Figure 6.5).

6.2.2 Ecological Status and Environmental Legislation

The European Community's Water Framework Directive (WFD; Anon, 2000) requires from the member states to aim at achieving "good" ecological status or maintaining "good" or "high" ecological status of its coastal waters by 2015. The trophic status of the Gulf of Riga has been clearly, but probably not very drastically, altered since pre-industrial years (Håkanson & Bryhn, 2008b) so the present trophic state may be classified into the category of "moderate" ecological status. This would mean that the goals in the WFD require substantial action to decrease the eutrophication in the Gulf. The jurisdiction of the WFD only covers parts of the Gulf of Riga as per the definition of coastal waters in WFD. However, the water policy was complemented in 2008 by the Marine Strategy Framework Directive (MSFD; Anon, 2008), which stipulates that EU member states should also take necessary measures to achieve (or maintain) good ecological status in marine waters by the year 2020. The MSFD applies to the whole Gulf of Riga which consists of Estonian and Latvian territorial waters (Figure 6.1).

Sewage treatment is an effective and cost-effective measure for decreasing P inputs to the Baltic Sea (Gren & Elofsson, 2008; Bryhn, 2009). P removal in sewage treatment plants has been expanding gradually since the 1970s (Kotta *et al.* 2008). Urban sewage emissions in the EU are regulated by the Urban Waste Water Treatment Directive. The latest summary report on implementation in 2006, only 7 out of 71 Latvian agglomerations which were connected to sewage treatment plants complied with the Directive's requirements (GHK, 2006). Table 6.1 shows which loading reductions to the Gulf of Riga would be expected compared to the year 2004 if sewage treatment would occur in the drainage area according to the Urban Waste Water Treatment Directive. As a comparison, the average waterborne P loading to the Gulf during 1997–2003 was 2,180 tonnes per year according to HELCOM (2007).

Table 6.1 Expected reductions (tonnes/year) compared to 2004 in total phosphorus reductions to the Gulf of Riga if sewage treatment in the drainage area would be adjusted according to the Urban Waste Water Treatment Directive. Data from HELCOM (2007).

Belarus	Estonia	Latvia	Lithuania	Russia	Total
523	17	162	46	431	1,179

6.2.3 Conflicts of interest

The quality of Baltic Sea waters have previously received very high non-market price tags in valuation studies where respondents were asked how much they were willing to pay to restore conditions prevailing before the 1960s. In the whole drainage basin of the Baltic Sea, the total willingness to pay for this goal was estimated at 3,600 million euro per year (Bryhn, 2009). Dampened algal blooms are likely to be welcomed by large sectors of the population around the Baltic Sea even if the undertaken measures should be financed by taxes. However, there are also potential conflicts of interest which may arise in connection with P abatement. For instance, decreased agricultural production is a potential measure against eutrophication which may be very unpopular in most rural areas. It should, however, be noted that there are other effective, cost-effective and probably less controversial abatement options available, such as improved urban sewage treatment (Bryhn, 2009).

There is also a potential conflict between proponents of nutrient abatement and stakeholders in the fisheries sector. The regression in Figure 6.4 gives at hand that a 10% decrease in primary production would result in a 9% decrease in fish yield. This would correspond to a decreased herring yield of about 9,000 tonnes per year according to Figure 6.5. The Baltic States are among the most severely affected countries in the world by the ongoing global recession. The Latvian economy shrunk by 18% in 2009 and unemployment exceeded 20% in February 2010 (IMF, 2010). It may be therefore be difficult to raise support for financing abatement efforts in these countries within the near future. Furthermore, the most cost-effective solution to the eutrophication problem in the Gulf and the Baltic Sea would include that a substantial share of the cost regarding nutrient abatement around the Gulf of Riga is carried by the Western European countries (Gren & Elofsson, 2008; Bryhn, 2009).

Table 6.1 shows that P reductions to the Gulf by sewage treatment have particularly great potential in Russia and Belarus, neither of which is an EU member. These two countries also have the lowest gross domestic product per capita (at purchase power parity) among the five countries which share the drainage basin of the Gulf of Riga (World Bank, 2010). However, the Gulf's coastline is shared by EU member states Estonia and Latvia (Figure 6.1). It may be difficult to motivate or enforce great and costly reductions in Russia and Belarus while Estonians and Latvians will reap much of the benefit. Nevertheless, improved sewage treatment in Russia and Belarus should in any case improve the quality and function of freshwater ecosystems in the Russian and Belarusian upper parts of the drainage basin.

6.3 THE PARTICIPATORY PROCESS: THE CITIZENS'S PANEL, SCIENTISTS AND POLICY MAKERS

In AWARE, the participatory process in the Gulf of Riga case study was formed during several events which aimed at increasing the connectivity between citizens, policy makers and scientists by means of workshops, a conference, interviews and online surveys. The structure of this participatory process has been described in Chapter 5. The weak part of the process was primarily the online survey which was unfortunately held during the Latvian and Estonian 2010 summer holiday season and therefore attracted only five respondents. Results from this survey will therefore not be commented in this chapter.

6.3.1 AWARE local workshop on the Gulf of Riga

6.3.1.1 Organisation

The workshop took place on October, 29–30, 2010 in Jurmala, Latvia. The scope of the workshop was on how to achieve a sustainable ecosystem in the Gulf of Riga. The focus was on the present status and pressures on the Gulf of Riga and elaboration of potential future scenarios. The Citizen's Panel met scientists (i.e. representatives from the AWARE partner group) and various local experts and managers who lectured about societal and environmental problems that the Panel earlier had identified as particularly important (the problems relevant for the Gulf of Riga were already identified during the European workshop held on 22–24 April, 2010 in Paris). In total, 23 participants took an active role in the workshop. The key target group was Citizen's Panel consisting of 10 selected local inhabitants of the Gulf of Riga (8 from Latvia and 2 from Estonia). Another target group was stakeholders representing agriculture, tourism, fishery and environmental sectors. The third group was scientists represented by the AWARE project team or Latvian researchers on the status of the Gulf of Riga. The various AWARE partners had different complementary roles during this workshop. Baltic Environmental Forum-Latvia was responsible for the organisation and moderation of the workshop.

6.3.1.2 Agenda

The local workshop was structured based on the DPSIR (Drivers–pressure state-impact- response) framework commonly used by for environmental assessments. During the session on the present situation in the Gulf of Riga, information on trophic status development (N&P concentrations, water clarity, Chlorophyll) and impacts on human health (bathing waters, food) and natural resources (algal blooms, fish) was presented. The aim was to feed the Citizen's Panel with facts and understanding on basics in eutrophication processes and related change of trophic status in coastal waters. The EU classification of water quality status (high, good, moderate, poor, bad) was introduced and relevant indicators to measure the status were presented by the scientists. This was followed by a facilitated discussion between the scientists and citizens on the lucidity of definitions and intelligibility of the evaluation of the water status. Upon a request of the Citizen's Panel, an issue of other pollutants such as oil spills and hazardous substances was included in the agenda. Scientists were explaining effects caused by these pollutants to environment and human health, as well as presenting facts on occurrence and risks for leakages of substances in the Baltic Sea. The session on the present situation in the Gulf of Riga was closed with a presentation on sources of the nutrient pollution to the Gulf of Riga. Point sources and related human behaviour, agricultural production and nutrient losses from different crops were particularly addressed by the speech and discussion.

The second session was devoted to development of future scenarios on the status of the Gulf of Riga. Several scenarios on pollution load reductions to achieve good environmental status of the Gulf of Riga was presented to the citizens by the scientists. It was followed by the group discussions just among the citizens on key certainties and uncertainties with regard to pollution loads and impacts to be considered. The key sectors like, agriculture, forestry, industry and households were reviewed.

The third sessions was dedicated to policy response to the environmental situation of the Baltic Sea. The representative of the Ministry of the Environment explained the national policies and legal acts adopted by the Parliament and Government of Latvia to ensure the protection of coastal and marine waters. It was followed by the discussion on the connecting policy ambitions on achieving good environmental status and involved costs to be bearable for the society. As the Gulf of Riga drainage basin is shared by non-EU countries the presentation on international cooperation with Russia and Belarus was contributing to the session on policy response.

A large part of the workshop was devoted to issues related to different measures to be taken by the different polluting sectors in order to achieve a better quality of waters and to use resources in sustainable way. The Citizen's Panel was informed about the environmental regulation regarding wastewater treatment plants and status of complying with this regulation in Latvia. The information on how centralised and individual sewage collection systems are planned and built about was provided by the Latvian Environmental investment Fund. Representatives from the Farmers Union and the Latvian Rural Advisory and Training Centre informed about trends in farming, measures taken to reduce nutrient pollution, and various capacity building activities targeted to the agricultural sector. Fishery is an economical sector with long historical traditions in the Gulf of Riga region. Information about main fish species and the landings and stocks trends of these species were provided by an expert from the Institute of Food Safety, Animal Health and Environment – "BIOR". It was pointed out that during the last years the number of fishing vessels has been reduced drastically, thus threatening the basis for existence of the Latvian fishery as an economy sector in the long term.

The last sector discussed during the workshop was tourism related to coastal area. The expert from Latvian Tourism development State Agency presented the national tourism strategy and stressed the importance for rural issues to include coastal tourism. During the thematic sessions different viewpoints and positions of the citizens were gathered. They formed a basis for elaboration of the Citizen's Statement on actions to achieve sustainable coastal management in the Gulf of Riga.

6.3.1.3 Citizens' statement

Having an intensive 2-day workshop with presentations and discussions among and between citizens, scientists and managers/implementers, a Citizens' Statement and adaptations of the scientific scenario building to the citizen's recommendations were elaborated. The Statement was elaborated by means of the following bullet-point list:

Classification of water status used by the WFD and MSFD

- The definitions of the current classifications categories (high, good, moderate, poor, bad) are rather general and need common and detailed interpretation. The percentages are necessary to understand what the indicators are saying, for example, 10% deviation from the "high" value means "good" status. This would also enhance comparability across Europe.
- Simple and accessible information on water quality based on key indicators (e.g. N, P, Secchi depth, chlorophyll) is a key for citizens to follow the situation on the status of the Gulf of Riga. Indicators shall also reflect information related to health risks.

Hazardous substances

- Public information is needed on the sources of all hazardous substances, thus to be aware how to avoid the use of certain products.
- Public dialogue about hazardous substances, but also about daily products for example detergents, fertilizers, and batteries in boat harbors, which all affect coastal waters shall be organised and increased involving policy makers, scientists and relevant stakeholders.
- As the present statistics reflects only larger oil spills, records and monitoring shall be improved for small scale oil spillage at the local level.
- Public information and communication about policy changes to consumer products shall be improved (e.g. why old light bulbs suddenly are banned).

Scenario-building

- One shall acknowledge that any implemented changes in pollution reduction will not come about quickly in the state of the environment.

- When developing different scenarios for setting the ecological goals human health risks shall be considered above economic needs. On the other hand ambitious ecological goals should be balanced with social needs and acceptable costs.
- Encourage gradual planning and actions towards improvement of natural resources
- Increase public awareness about the trade-off in costs now vs. costs later for the future water quality
- Increase the efficiency of urban waste water treatment.
- To achieve more cost-efficient treatment of waste water, consider investing nationally and across borders.
- Increase cooperation among scientists across borders.

Pollution loads and impacts

- Policy makers and scientists shall provide citizens with clear and simple information on different alternatives (e.g. on organic agriculture).
- The work shall be focus also on an increase in industrial transparency (e.g. publish sewage treatment results) and explanations of impacts on human health.
- To achieve reduction in pollution load caused by consumption, comprehensive information on the health and environmental impacts of consumer products shall be provided regularly. Communication on "consumer rights" regarding environmental effects of consumer products shall be increased.

Policy

- Before deciding on policy measures, evaluate their cost effectiveness.
- As the status of the Gulf of Riga depends on neighbouring countries located in drainage basin but not having direct marine waters, they need financial incentives to cooperate for achieving the good environmental status in the Baltic Sea.
- Necessary resources for monitoring, enforcement, and controlling on human and environmental impacts shall be maintained despite of the economic crises.
- Enforcements and monitoring measures should be supported by economic instruments/sanctions/penalties.
- Awareness on water pollution and quality status, especially in rural areas, shall be raised, for example, offer free water quality tests for individual wells.
- Scientific ambassadors who could explain/interpret investigation results to policy makers and to the public shall be employed. Involve more scientists in environmental assessments and also in public hearings.
- Clear information about impacts of policies (on people's daily lives) is needed, and appropriate involvement of stakeholders in the decision-making should follow.
- Water policies should be communicated in a more proactive way using a variety of media for public communication (e.g. official announcements and advertisements in local newspapers; regular email lists; TV evening news; TV panel discussions; radio; popular online public portals; social networks).
- Integrate "water days" into existing events, open to all, where achievements can be presented and awareness raised.
- Encourage a common agricultural strategy at the national level.
- Improve the communication between farmers, agricultural and rural development advisors, and the public, especially on best practices.
- For tourism, check the carrying capacity of the local area, and consider long-term maintenance of tourism infrastructure.

6.3.2 The local public conference

The scientific approach for the local conference compared to the approach at the local workshop was mainly to adjust scenarios with the aim to meet suggestions made by the Citizens' Panel. There were less than two days between the end of the local workshop and the beginning of the local conference, but some adjustments to scenarios according to the Citizens' Statement could nevertheless be done. The existing scenarios had already showed and accounted for that any implemented changes in pollution reduction or improved water quality will not come about quickly, so this suggestion was already considered to be met. Using cost-effective measures and international cooperation was also already used and discussed in scenario contexts. Referring to abatement costs in euro per capita instead of the total cost in euro was accepted and this only required a few simple calculations.

To "balance ambitious ecological goals with social needs and acceptable costs" or so called "the golden middle way" remained the main challenge for scenario building. Even if full action would be taken, the water quality of the Gulf cannot reach "good" ecological status in 2020 as required in the EU Marine Strategy Framework Directive. The previous focus was on an ambitious nutrient reduction programme which would eventually lead to "good" ecological status and would cost 87–220 euro per drainage basin inhabitant and year. This scenario, compared to the "no action" alternative is described in the figure below.

The new focus was instead put on a less ambitious action plan, to implement the EU Urban Waste Water Treatment Directive in the whole drainage basin of the Gulf. This option would cost 9–20 euro per capita and year, which was seen as an acceptable cost and a "golden middle way" by the Citizens' Panel.

6.3.2.1 Organisation

The local conference was held on November, 1, 2010, at the Latvian Ministry of Environment in Riga, Latvia. The theme of the conference was the same as for the workshop, that is, how to achieve a sustainable ecosystem in the Gulf of Riga. The unique objective of the conference was to present and discuss further cooperation needs and participatory approaches between citizens, scientists and other stakeholders in the field of water management, particularly in integrated water and coastal zone management. The conference aimed at focusing on the perceptions of the possible futures of the Gulf of Riga as well as to elaborate the recommendations for enhancement of the management measures towards improvement of the status of the coastal and marine waters. The various AWARE partners had different complementary roles during the conference. BEF Latvia was responsible for organising the conference and chairing the event. Uppsala University gave two presentations on environmental problems in the Gulf and constructed an updated scenario regarding recovery from eutrophication. Bioforsk moderated the panel discussion. Adelphi conducted the evaluation. Additionally, some speakers from the relevant Latvian and Estonian authorities were invited. Those invited to the conference were the AWARE Gulf of Riga partners, the Citizen's panel and identified stakeholders which represent either international organisations, national government (Ministries of the Environment of Latvia and Estonia, Ministry of Transport, Ministry of the Regional Development and Local Governments, Ministry of Agriculture, Fishery Department), municipalities, scientific and research institutions, NGOs, training institutions, environmental and nature NGOs, consultants. The number of registered participants was 56 while 49 people actually participated.

6.3.2.2 Content

The agenda of the local conference was structured to dedicate most of the time to future of the Gulf of Riga and develop on scenarios. The first content related session of the conference gave an overview on the current policy framework on coastal and marine protection in the Baltic Sea and the Gulf of Riga. Representatives of the Estonian and Latvian Ministry of the Environment told about the existing legislative requirements, environmental targets and implications for sectors and estimated funding needed to achieve the targets. Further, the Swedish and Latvian scientists presented jointly the status of the Gulf of Riga. According to the HELCOM Assessment Report 2010, the status is not better than moderate and eutrophication is one of the key problems of the Gulf of Riga. This introductory session on the present status was followed by the presentation of two members of the Citizens' Panel who introduced to the Statement elaborated during Jurmala workshop on 29–30 October. The recommendations were also handed out to participants of the local conference. As the next, scientists presented two scenarios for achieving good environmental status in the Gulf of Riga. The AWARE project partner from the Uppsala University presented a scenario based on the introduction of the urban waste water treatment with P removal in the whole drainage basin (see Figures 6.6 and 6.7). The researcher at the Latvian Institute of Aquatic Ecology presented a change in the status of the Gulf of Riga if the pollution load will be reduced according to the HELCOM targets (Table 6.1). While the second presentation addressed the development trends of key indicators, the AWARE project partner additionally presented also the calculated costs for achieving the scenarios.

The last session of the conference was organised as a panel discussion related to the future of the Gulf of Riga and feedback on the Citizen's Statement. The Panel consisted of 2 scientists, 2 policy makers and 2 members of Citizen's Panel. Two topics were discussed: pollution prevention measures, and interaction between science, citizens, and policy making.

6.3.2.3 Main conclusions from the conference

The general impression of this conference was that each section; scientists, policy makers and citizens were able to debate based on the same terms and increase a better connectivity with each other. The citizens stressed that information should be easy to understand and especially that every action has to be motivated and accessible for every involved section for example citizens and stakeholders. Citizens want to reach a better quality in the environment on a step-by-step basis so the goal seems more approachable. Inclusion of health risk was stressed by the citizens. The economic costs could instead of the whole sum be presented as a cost per capita to maintain a better substance of the actual costs. More scientists should be involved in the process of decision making and a "scientific ambassador" would make it easier to explain information to policy makers and the public. A better communication between different sections would increase the awareness and lead to actions that everyone understand the pros and cons of. The connectivity between scientists, policy makers and citizens were especially challenged under the panel discussion and each part was able to understand each other better when they were forced to use "the same language" to keep the discussion alive.

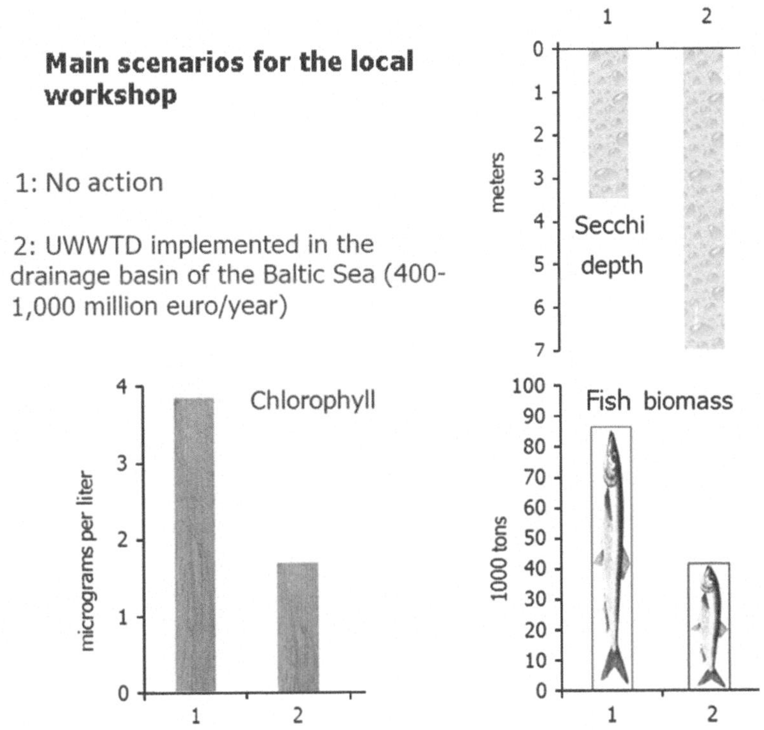

Figure 6.6 Initial scenarios to achieve good environmental status in the Gulf of Riga. (UWWTD is the EU Urban Waste Water Treatment Directive. Chlorophyll is a phytoplankton pigment which is used as a proxy variable for phytoplankton ("algal") bloom intensity. Secchi depth describes the water clarity; and specifically at which maximum depth a white disc is visible from the surface).

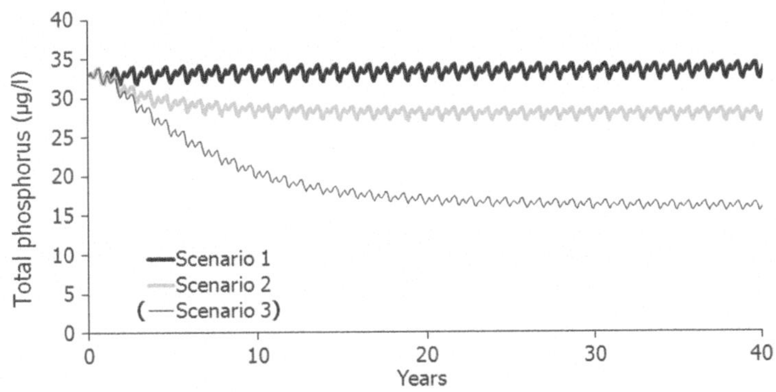

Figure 6.7 The scenarios for total P concentrations in the Gulf of Riga. (Scenario 1: no action; Scenario 2: UWWTD is implemented in the drainage basin of the Gulf of Riga; Scenario 3: UWWTD is implemented in the whole drainage basin).

6.4 INTERVIEWS WITH KEY STAKEHOLDERS AND POLICY MAKERS

6.4.1 Organisation

The purpose of the AWARE case study interviews with key stakeholders and policy makers was to inform about and gather input on the AWARE approach; to find out how policy makers perceive the results of the local workshop/conferences (i.e. the citizens recommendations); and to find out to what extent the interviewees understand and take into account the citizen recommendations developed in the local events.

The interviews were pretested and changed according to comments and the partners developed a guideline for interviews. The face to face or telephone interviews were conducted in the period of March–April 2011. Ten policy makers from Estonia and Latvia were interviewed. The respondents were asked six questions regarding the AWARE approach, citizen recommendations and how scientists, policy makers and citizens can cooperate. The interview respondents were representatives from the Latvian Ministry of Agriculture (policy makers from Fishery and Agriculture sectors), the Latvian Environment, Geology and Meteorology Centre, the Latvian Ministry of Environmental Protection and Regional Development (three experts on marine water policy, river basin management including coastal waters, terrestrial coastal zone management), the Estonian Ministry of Environment and the Peipsi Centre for Transboundary Cooperation (an international non-profit institute, which is part of the working group in the Estonian-Russian Transboundary Water Commission.

Unfortunately it was not possible to interview as many policy makers from Estonia, as from Latvia. An observation made by the interview team was that it was generally difficult to get acceptance to conduct an interview even though it was informed that the interview time was only 30–45 minutes and comprehensive information was delivered about the subject both in text and orally prior to the interview. Lack of time (i.e. busy with other duties) and that the subject was not their expertise area was mentioned as explanations as well as some felt language difficulties (in Estonia since the interviews were conducted in English). The latter was surprising since the interview objects spoke good English during the initial contact.

6.4.2 Main conclusions from interviews

The stakeholders and policy makers interviewed from the Gulf of Riga were generally very interested in the AWARE approach and its relevance for ongoing work on water issues. A few of the respondents did not have experience from similar approaches and were a bit more sceptical to the approach and its relevance for their work. The policy makers generally seemed interested in developing the cooperation between scientists, policy makers and citizens. They commented that the approach is possible for long-term-sustainable development planning, but that it will be more challenging to use in managing the daily, emerging issues. The policy makers expressed that the success of implementing the AWARE approach would depend very much on the availability of human and financial resources. Another issue is how to improve communication and cooperation between the three groups (citizens, scientists & policy makers).

When asked if they would expect any impacts on coastal waters quality due to the AWARE approach, the majority of respondents answered that they expected impacts in the long run. Short term impacts were cited as more unlikely as it takes time for the environment to respond. It was commented that an integrated approach will have a long-term effect since it will lead to less misunderstanding and that inclusion of all relevant stakeholders will create better acceptance in relation to decisions and that one might avoid rapid changes in policy over the elected mandate periods if there is a well-established consensus. The policy makers responded well to the set of recommendations given by the citizens, regarding the management of coastal waters. The majority of the respondents had heard similar citizen concerns before. The new aspects given in the recommendations were related to the involvement of scientists in the process (not just the citizens and policy makers) and that citizens acknowledge that the improvement of the environmental situation is slow process. Public participation is already somewhat included in the management system but there are existing difficulties with implementing it on a broader scale. A concern among some of the respondents was on the citizen selection process and the subjectivity of the engaged persons. The process of selecting citizens should be done with care, making sure that some groups do not dominate the process.

The majority of the respondents was positive to integrated adaptive water management and said it was applicable in practice. The respondents had experiences with similar approaches, such as in relation to public hearings and special coordination groups, and gave several examples of best practices particularly highlighting existing river basin cooperation. The most pressing management aspects for environmental degradation of the coastal waters in the Gulf of Riga, according to the policy makers were: lack of enforcement of regulation/ policies and lack of sufficient funding. The majority of respondents agreed in with all the proposed topics and felt that they were closely linked. Environmental awareness was also considered very important.

It was clear from the interviews that more information sharing is important, both between scientists and policy makers/citizens and between policy makers and the public. The environmental threats seem to be an issue that needs more debate and more information to all stakeholders. There was a wide consensus among the respondents that eutrophication is a serious threat. Decreasing abundance of fish as a threat was considered more difficult to comment on for most of the respondents and this can be due to lack of information and possibly cooperation between sectoral ministries. The policy makers had several suggestions on what they can do to improve communication between decision makers and the public in their daily work, such as: participate more actively in the events organised by different stakeholders, send information to press or ongoing TV programmes on environmental issues, improve communication via newspaper articles and short reports in an easily understandable format and be clearer in explaining impacts and discuss the pros and cons in a more accessible way for all groups. It was also advised that scientists need to participate more actively in some processes and present information and results from research in an understandable way to policy makers and citizens. It was stated by respondents that the integration and connectivity between policy-science and stakeholder is needed and there should be more meeting arenas. There should also be more meeting arenas between sectoral ministries and not only competition between the various ministries.

The respondents were asked what they think is the role of citizens, scientists and policy-makers in addressing the potential environmental threats. The common view among respondents was that policy makers generally shall make decisions based on scientific evidence and in line with the opinions of the society. Policy makers are acting according to available information, however, sometimes information is lacking to prepare good quality policy documents. Citizens are responsible in terms of environmental awareness and behaviour. The representatives of policy makers in the Gulf of Riga generally agreed that the approach of AWARE is in line with relevant policy frameworks and they were positive to the approach. However, financial and human resources are a challenge in the implementation of more cooperation between scientists, policy makers and citizens.

6.5 CONCLUSIONS AND RECOMMENDATIONS FROM THE GULF OF RIGA CASE STUDY

The Gulf of Riga case study increased and improved the communication and connectivity between citizens, scientists and policy makers. Particularly between the first two groups stronger relationships were set up as the scientists were also key partners of the projects and implementing all project activities not only targeted to the Gulf of Riga. More effort could have been made at attracting more policy makers to the process. They turned out to appear as receivers and feedback givers. However, the workshop, the conference and the interviews were performed successfully and stimulated many discussions between the three groups, which indicates that the overall outcome of the case study was good.

The participants from all three groups expressed gratitude and content towards the AWARE project organisers; apparently it was meaningful for group participants to meet, discuss, get new viewpoints and improve their knowledge. Citizens gradually became more knowledgeable, losing their roles as "ordinary" citizens and gaining new roles somewhere between layman ship and expertise. Policy makers appreciated concern for the decisions they make. Overall, it was expressed that a participatory process such as the AWARE approach may serve as a complement to current science and policy making. The AWARE process was viewed as a means for democratization of science and policy in the European Union; as a means for decreasing the height of the "ivory towers" that scientists and policy makers are sometimes considered to sit and act on top of.

REFERENCES

Anon. (1991). Council Directive 91/271/EEC Of 21 May 1991 Concerning Urban Waste-Water Treatment. *Official Journal of The European Communities*, **L135**, 40–52.

Anon. (2000). Directive 2000/60/EC of The European Parliament and of The Council of 23 October 2000 Establishing A Framework For Community Action In The Field of Water Policy. *Official Journal of The European Communities*, **L327**, 1–72.

Anon. (2008). Directive 2008/56/EC of The European Parliament and of The Council of 17 June 2008 Establishing A Framework For Community Action In The Field of Marine Environmental Policy. *Official Journal of The European Communities*, **L164**, 19–40.

Bryhn A. C. (2009). Sustainable phosphorus loadings from effective and cost-effective phosphorus management around the Baltic Sea. *PLoS ONE*, **4**, e5417.

Bryhn A. C. (2010). Trends in total phosphorus loadings and concentrations regarding surface waters of the Baltic Sea, 1968–2007. *The Open Oceanography Journal*, **4**, 1–8.

Bryhn A. C. (2012). Estimated trophic state effects and abatement costs in connection with improved Urban Sewage Treatment in the Gulf of Riga. *Journal of Environmental Engineering*, in press.

Bryhn A. C. and Håkanson L. (2010). Land uplift effects on the phosphorus cycle of the Baltic Sea. *Environmental Earth Sciences*, **62**, 1761–1770.

Chassot E., Melin F., Le Pape O. and Gascuel D. (2007). Bottom-up control regulates fisheries production at the scale of eco-regions in European Seas. *Marine Ecology Progress Series*, **343**, 45–55.

Christensen V. and Walters C. J. (2004). Ecopath with ecosim: methods, capabilities and limitations. *Ecological Modelling*, **172**, 109–139.

EC-JRC (2008). Chlorophyll-A Concentrations, Temporal Variations and Regional Differences From Satellite Remote Sensing. HELCOM Indicator Fact Sheets 2009. http://www.helcom.fi/environment2/ifs/ifs2008/en_GB/chlorophyll/ (accessed 17 February2012).

Enell M. and Fejes J. (1995). The Nitrogen load to the Baltic Sea – present situation, acceptable future load and suggested source reduction. *Water, Air, & Soil Pollution*, **85**, 877–882.

Eronen M., Glückert G., Hatakka L., van de Plassche O., van der Plicht J. and Rantala P. (2001). Rates of Holocene isostatic uplift and relative sea-level lowering of the baltic in SW Finland based on studies of isolation contacts. *Boreas*, **30**, 17–30.

EUROSTAT (2011). Annual Average Rate of Change In Harmonized Indices of Consumer Prices. http://epp.eurostat.ec.europa.eu (accessed 2 June 2011).

Fleming-Lehtinen V. and Kaartokallio H. (2009). Water Transparency In The Baltic Sea Between 1903 and 2009. HELCOM Indicator Fact Sheets 2009. http://www.helcom.fi/environment2/ifs/ifs2009/secchi/en_GB/WaterTransparency%20/ (accessed 17 February 2012).

Gren I.-M. and Elofsson K. (2008). Costs and Benefits From Nutrient Reductions To The Baltic Sea. Swedish EPA Report 5877, Stockholm, p67 p.

GHK (2006). National Evaluation Report For Latvia, Executive Summary, GHK, Brussels, p. 18.

Håkanson L. (2009). A general process-based mass-balance model for phosphorus/Eutrophication as a tool to estimate historical reference values for key bioindicators, as exemplified using data for the Gulf of Riga. *Ecological Modelling*, **220**, 226–244.
Håkanson L. and Boulion V. V. (2002). The Lake Foodweb – Modelling Predation and Abiotic/Biotic Interactions, Backhuys, Leiden, p. 344.
Håkanson L. and Bryhn A. C. (2008a). Eutrophication In The Baltic Sea – Present Situation, Nutrient Transport Processes, Remedial Strategies. Springer, Berlin, p. 261.
Håkanson L. and Bryhn A. C. (2008b). Tools and Criteria For Sustainable Coastal Ecosystem Management, Examples from the Baltic Sea and Other Aquatic Systems. Springer, Berlin, p. 292.
Håkanson L. and Bryhn A. C. (2008c). Modeling the foodweb in coastal areas: a case study of RingkøBing Fjord, Denmark. *Ecological Research*, **23**, 421–444.
Håkanson L., Bryhn A. C. and Eklund J. M. (2007). Modelling phosphorus and suspended particulate matter in RingkøBing Fjord in order to understand regime shifts. *Journal of Marine Systems*, **68**, 65–90.
Håkanson L. and Eklund J. M. (2007). A dynamic mass balance model for phosphorus fluxes and concentrations in coastal areas. *Ecological Research*, **22**, 296–320.
Håkanson L., Ragnarsson Stabo H. and Bryhn A. C. (2010). The Fish Production Potential of The Baltic Sea. Springer, Berlin, p. 340.
Hansson S., Hjerne O., Harvey C., Kitchell J. F., Cox S. P. and Essington T. E. (2007). Managing Baltic Sea fisheries under contrasting production and predation regimes: Ecosystem model analyses. *Ambio*, **36**, 265–271.
Hassler B. (2003). Science and Politics of Foreign Aid. Swedish Environmental Support to the Baltic States. Kluwer Academic Publishers, Dordrecht, p. 221.
HELCOM (2007). Towards A Baltic Sea Unaffected By Eutrophication, HELCOM, Krakow, p. 35.
HELCOM and NEFCO (2007). Economic Analysis of The BSAP with Focus on Eutrophication, HELCOM, Krakow, p. 112.
Houde E. D. and Rutherford E. S. (1993). Recent trends in estuarine fisheries: prediction of fish production and yield. *Estuaries and Coasts*, **16**, 161–176.
ICES (2008). ICES Advice 2008, Book 8, ICES, Copenhagen, p. 77.
IMF (2010). After Severe Recession, Stabilization In Latvia. IMF Survey Online. http://www.imf.org/external/pubs/ft/survey/so/2010/car021810a.htm (accessed 23 February 2010).
Iverson R. L. (1990). Control of marine fish production. *Limnology and Oceanography*, **35**, 1593–1604.
Jennings S. and Brander K. (2010). Predicting the effects of climate change on marine communities and the consequences for fisheries. *Journal of Marine Systems*, **79**, 418–426.
Jonsson P., Carman R. and Wulff F. (1990). Laminated sediments in the Baltic – a tool for evaluating nutrient mass balances. *Ambio*, **19**, 152–158.
Kotta J., Lauringson V., Martin G., Simm M., Kotta I., Herkül K. and Ojaveer H. (2008). Gulf of Riga and PäRnu Bay. In: Ecology of Baltic Coastal Waters, U. Schiewer (ed.)., Springer, Berlin, pp. 217–243.
Laznik M., Stålnacke P., Grimvall A. and Wittgren H. B. (1999). Riverine input nutrients to the Gulf of Riga – temporal an spatial variation. *Journal of Marine Systems*, **23**, 11–25.
Looring M. (2009). The impact of population density and urbanization in the coastal northeast regions of the Baltic Sea: A time-spatial description. In: Integrated Coastal Zone Management, E. Moksness, E. Dahl and J. G. Støttrup (eds.)., Wiley-Blackwell, Hoboken, NJ, pp. 145–154.
Neumann T. and Schernewski G. (2008). Eutrophication in the Baltic Sea and shifts in Nitrogen Fixation analyzed with a 3D ecosystem model. *Journal of Marine Systems*, **74**, 592–602.
Nixon S. W. (1982). Nutrient dynamics, primary production and fisheries yields of Lagoons. *Oceanologica Acta*, **4**(Suppl.), 357–372.
Ojaveer H., Lankov A., Eero M., Kotta J., Kotta I. and Lumberg A. (1999). Changes in the ecosystem of the Gulf of Riga From The 1970S to the 1990S. *ICES Journal of Marine Science*, **56**(suppl.), 33–40.
Savchuk O. P. (2005). Resolving the Baltic Sea into seven subbasins: N and P budgets for 1991–1999. *Journal of Marine Systems*, **26**, 1–15.
Savchuk O. P. and Wulff F. (2009). Long-term modeling of large-scale nutrient cycles in the entire Baltic Sea. *Hydrobiologia*, **629**, 209–224.
Schindler D. W., Hecky R. E., Findlay D. L., Stainton M. P., Parker B. R., Paterson M. J., Beaty K. G., Lyng M. and Kasian S. E. M. (2008). Eutrophication of lakes cannot be controlled by reducing Nitrogen input: results of a 37-year whole-ecosystem experiment. *PNAS*, **105**, 11039–11040.
Stålnacke P., Tamminen T., Vagstad N., Wassmann P., Loigu E. and Jansons V. (1999). Nutrient runoff and transfer from land and rivers to the Gulf of Riga. *Hydrobiologia*, **410**, 103–110.
Tyrrell T. (1999). The relative influences of Nitrogen and Phosphorus on oceanic primary production. *Nature*, **400**, 525–531.
Ware D. M. and Thomson R. E. (2005). Bottom-up ecosystem trophic dynamics determine fish production in the north-east pacific. *Science*, **308**, 1280–1284.
World Bank (2010). World Development Indicators Online. http://publications.worldbank.org/WDI/ (accessed 15 March 2012).

ANNEX: MODELLING STRATEGY

This section describes the basics of the nutrient, fish, chlorophyll, Secchi depth and economic cost modelling strategy for the Gulf of Riga. A more thorough motivation is provided in Bryhn (2012). The limiting nutrient in primary production of the Gulf of Riga is P (Kotta *et al.* 2008), so a crucial task in predicting effects on the trophic state from abatement action is to accurately assess external and internal P fluxes to, from and within the Gulf. If P concentrations would decrease gradually in the surface waters of the Gulf, a gradually aggravated P deficiency could be expected for the phytoplankton communities which would thereby be forced to decrease their activity, reproduction and biomass (Tyrrell 1999; HELCOM, 2007; Håkanson & Bryhn 2008a; Schindler *et al.* 2008).

Mass-balance models are the only means by which changes in trophic state and internal P fluxes can be predicted from changes in P loading and in other external factors (Håkanson *et al.* 2010). The model used in this study is called CoastMab and consists of mass-balance submodels of salt, TP and suspended particulate matter (SPM). Masses, concentrations and fluxes are simulated using monthly time steps and ordinary differential equations. There are also other sub-models connected to these mass-balance submodels; for example, a Secchi depth submodel, a chlorophyll submodel and also a foodweb model framework (CoastWeb) with predator-prey interactions (Håkanson *et al.* 2010). For simplicity and because no foodweb interactions were simulated in the Gulf of Riga case study, total fish biomass changes were not estimated by means of foodweb modelling but were instead predicted from TP concentrations and salinities. Thus, the CoastMab model (published in Håkanson *et al.* 2010 and further described in Bryhn, 2010) was used for predicting TP concentrations, chlorophyll-a concentrations (Chl; in µg/l), the Secchi depth (in m) and total fish biomass. Predictions of present levels of TP, Chl and Secchi depth were compared to empirical measurements from the current stationary trophic state described in Håkanson (2009).

Modelling changes in average subbasin conditions on longer (monthly, annual and decadal) timescales should be sufficient for the purpose of the investigations in this case-study and the CoastMab approach (Figure 6.A1) was therefore selected.

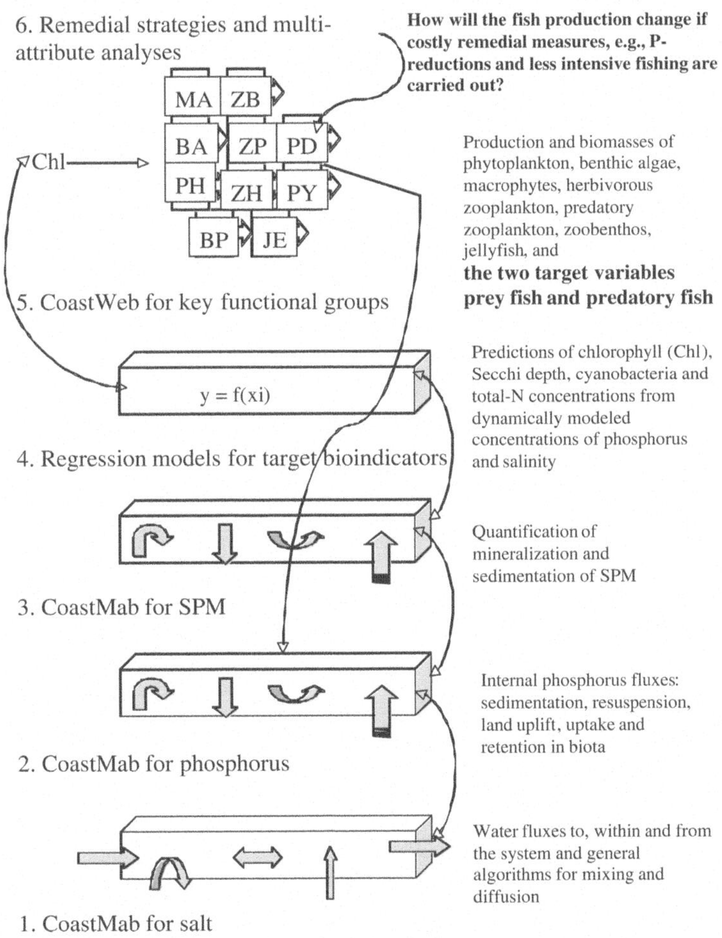

Figure 6.A1 A general outline of the structure of the CoastMab/CoastWeb-model.

TP and SPM fluxes and masses in the Gulf of Riga are thus simulated by CoastMab in the same manner as two other subbasins, the Bothnian Sea and the Bothnian Bay, using four state variables which represent surface waters, deep waters, ET sediments and A sediments. Six state variables are used for the Baltic Proper and the Gulf of Finland: surface waters and ET sediments, both above D_{WB}; middle waters and mid-range A sediments, both located between D_{WB} and the halocline; and, finally, deep waters and deep A sediments located below the halocline.

Some major simulated TP and SPM fluxes in CoastMab are sedimentation, mixing, resuspension (including erosion), diffusion, burial, biouptake and outflow. Sedimentation is the downward flux of SPM and particulate P due to gravity. Mixing is defined as the wind and wave driven TP and SPM fluxes between surface waters and lower water layers.

One flux category included in CoastMab but not in most other Baltic Sea nutrient or foodweb models is related to rising land and sediment areas. The isostatic land uplift after the last glaciation period occurs at a rate of 0–9 mm/year in the Baltic Sea region (Eronen *et al.* 2001). Land uplift gradually lowers the wave base and exposes new bottom areas to increased wind and wave driven erosion and resuspension. This means that large amounts of TP and SPM are continuously added to the system as a result of geological events (Jonsson *et al.* 1990). The eroded material primarily consists of glacial and postglacial clay, and the clay particles appear to have a clarifying effect on the Baltic Sea water. This effect has been used in the CoastMab model for improving the quantitative explanation to why waters are clear and oligotrophic in the northern Baltic Sea where land uplift is high while nutrient concentrations are much higher in the south where land uplift is low. Water mixing between subbasins is intensive, so geographical variations in nutrient loadings from the catchment cannot alone be the reason for the strong nutrient gradient in the Baltic Sea (Håkanson & Bryhn 2008a; Bryhn & Håkanson 2010).

Chl is tightly correlated to TP and total nitrogen concentrations in aquatic systems. Furthermore, Chl increases at increasing temperatures, while salinity in the surface water layer (Sal_{SW}, in psu) affects the Chl/TP ratio (Håkanson & Bryhn 2008a; Håkanson *et al.* 2010). In CoastMab, Chl is predicted from monthly surface water temperatures (SWT, in °C; the annual average and seasonal variation of SWT is modelled from the mean latitude of each subbasin), long-term mean salinities and dynamically modelled TP concentrations in surface waters.

Predictions from CoastMab have previously been successfully tested against long-term TP, Chl and Secchi depth data from the Baltic Sea subbasins displayed in Figure 6.1 (Håkanson & Bryhn 2008a), against long-term TP data from 21 Baltic Sea coastal areas (Håkanson & Eklund 2007) and against a long time series of TP, Chl and Secchi depth data in Ringkøbing Fjord, Denmark (Håkanson *et al.* 2007). The model error was in general smaller than the relative deviation in a comparison of empirical data divided into two randomly selected datasets. Fish biomass data are, however, comparatively uncertain and difficult to predict; empirical values in Ringkøbing Fjord were predicted in the right order of magnitude by Håkanson & Bryhn (2008c) using the CoastMab/CoastWeb approach.

The cost for improving urban sewage treatment to UWWTD standards was taken from HELCOM & NEFCO (2007) and was provided in 2004 euros. Costs were converted to 2011 prices by adjusting for inflation according to rates reported by EUROSTAT (2011). A separate cost estimate for the Gulf of Riga drainage basin was made using inflation adjusted costs (see Bryhn, 2012 for details).

Chapter 7

The diseased southern North Sea: current status and possible solutions

Josette Garnier[1,2], Paul Passy[1], Véronique Rousseau[3], Nathalie Gypens[3], Julie Callens[1], Marie Silvestre[1,2], Gilles Billen[1,2] and Christiane Lancelot[3]

[1]University Pierre and Marie Curie, UMR 7619 Sisyphe, boite 123, 4 place Jussieu, 75005 Paris, France
[2]CNRS, UMR 7619 Sisyphe & FR-3020, boite 123, 4 place Jussieu, 75005 Paris, France
[3]Ecologie des Systèmes Aquatiques (ESA), Université Libre de Bruxelles, Campus Plaine, CP 221, 1050 Brussels, Belgium

7.1 THE LAND TO SEA AQUATIC CONTINUUM

Over the past few decades, increased nutrient delivery by large rivers to the coastal waters has caused severe eutrophication problems in many places of the world's coastal ocean (Marchetti, 1991; Justic, 1991; Justic *et al.* 1995a,b; Diaz, 2001; Lancelot *et al.* 2007; Howarth, 2008; Glibert *et al.* 2008; Diaz & Rosenberg, 2008; Lancelot *et al.* 2011, among many others). The eutrophication-related symptoms may be quite diverse, depending on local physiographical and hydrological conditions of the coastal systems (Rabouille *et al.* 2008).

The interest in river eutrophication is relatively recent, compared to lakes and ponds (Vollenweider 1968), because rivers have been seen as systems capable of transporting, purifying and eventually evacuating all the pollutants discharged into them downstream. Similarly, the link between the coastal eutrophication and nutrient loads from the watershed basins was reported in the early 1980s (Officer & Ryther, 1980; Lancelot *et al.* 1987; Smayda, 1990; Cloern, 2001), but only recently quantified by a modelling approach from head waters to the sea (Billen *et al.* 2001, Cugier *et al.* 2005; Lancelot *et al.* 2007, Even *et al.* 2007).

In the coastal zone of the eastern English Channel and southern North Sea (French Coastal Zone [FCZ] and Belgian Coastal Zone [BCZ]), massive blooms of mucilaginous, unpalatable colonies of the haptophyte microalgae *Phaeocystis globosa* have been reported every spring since the 1970s (Lancelot *et al.* 1987; Lancelot *et al.* 2005; Lancelot *et al.* 2007; Cadée & Hegeman, 2002) and have been related to excess nitrate (Rousseau, 2000; Lancelot *et al.* 2009; 2011). Simultaneously, major changes in the deliveries of nutrients from the adjacent watersheds and in their ratio were discussed in terms of possible impacts at the coast (Billen *et al.* 2005; Billen *et al.* 2007; Thieu *et al.* 2009). The mean residual circulation of the water masses is such that nutrients delivered by the Seine, the Somme and the Scheldt rivers (the 3S, Table 7.1) cumulate along a SW-NE direction (Figure 7.1). The agricultural, domestic and industrial activities throughout the hydrological basin of these three rivers (i.e. the area they drain) result in very high nutrient loadings, particularly in terms of nitrogen and phosphorus. The system under study, that is the territory where management measures can be taken, thus includes an area as vast as 100,000 km^2 in France and Belgium.

Over this large territory, many other water issues are at stake, which should also be considered in water management policies. The area of concern thus includes the whole aquatic continuum from the headwaters of the Seine, Somme and Scheldt watersheds, their groundwater resources, the large rivers and their estuaries, and finally the coastal zones of the North of France and Belgium.

Table 7.1 Major characteristics of the 3S watersheds.

Basins	Area (km^2)	Pop. Dens ($inh.km^{-2}$)	Drain dens. ($m.km^{-2}$)	Mean Q ($m^3.s^{-1}$)	Urban (%)	Arable land (%)	Pasture (%)	Forest (%)
Seine	76 260	215	305	525	7	52	10	26
Somme	6190	110	115	35	6	77	19	8
Scheldt	19 920	540	164	140	25	39	25	8

Note: Pop. Dens.: population density; Drain Dens.: drainage density; Mean Q: mean discharge. % of land use in the basins.

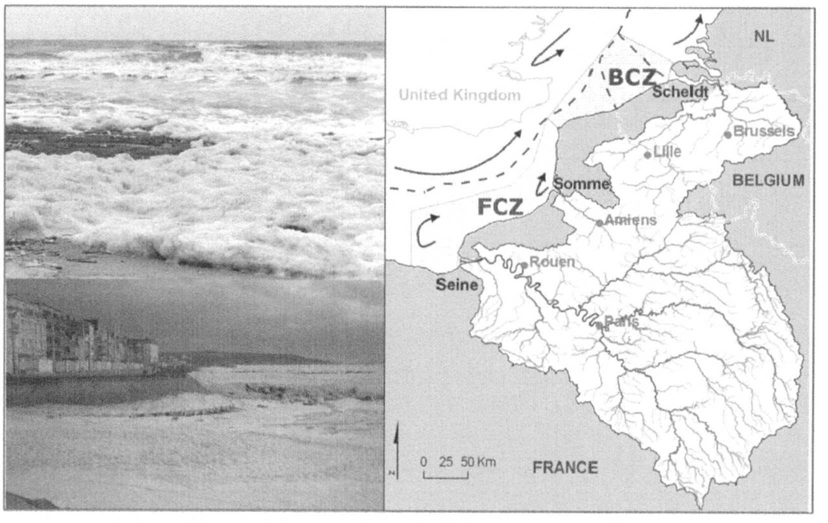

Figure 7.1 Left: Phaeocystis-related foam accumulation on a Belgian beach in spring (Photographs V. Rousseau). Right: map showing the French (FCZ) and Belgian (BCZ) Economic Exclusive Zones (dotted line) and the contributing 3S watersheds.

In addition to possible metal and organic micropollutants, rivers mostly deliver biogenic elements such as nitrogen (N), phosphorus (P) and silicon (Si) to the coastal sea, stimulating phytoplankton growth when light is sufficient. North Sea coastal eutrophication is the consequence of the increased river inputs of N and/or P originating from human activities in the watersheds, while silicon is little affected. Both the amount of nutrients delivered by rivers and their respective ratios determine the response of coastal ecosystems. All phytoplankton species require both N and P to grow, while only diatoms, the 'preferred' algae that feed zooplankton and fish, require Si. As long as N and P are brought by rivers in balance with Si, diatoms bloom, with beneficial effects for the food chain and the production of marine resources. On the contrary, when N and P are delivered in excess over Si, non-siliceous microalgae dominate the phytoplankton community. This opportunistic phytoplankton forms undesirable blooms often associated with harmful effects, such as, in the case of the Belgian coastal zone, foam accumulation (Lancelot *et al.* 1995; Rousseau *et al.* 2000; Figure 7.1).

Based on the river continuum concept (RCC; Vannote *et al.* 1980), the river constitutes a succession of interdependent systems, where biogeochemical processes transform nutrients released from the watershed during their transport from the headwaters to the coastal zone so that there is no direct link between nutrient emissions to the river and export to the sea. Indeed, throughout the hydrographic network and in its stagnant bodies such as ponds and reservoirs, processes of retention or elimination of nutrients modify the fluxes that are transferred downstream. The riparian zones that form the interface between the soil of the watershed basin and the surface water act as very efficient filters for nitrates of agricultural origin (Howarth *et al.* 1996; Tilman *et al.* 2001; Sebilo *et al.* 2005). Since N, P and Si have different sources and are involved in different mechanisms, the ratios between these elements change considerably from the upstream to the downstream sectors of the hydrographical network. These ratios are efficiently compared to the phytoplankton cell requirement. The mean N:P ratio of aquatic plants is about 7 (weight:weight; Redfield *et al.* 1963); the mean Si:N ratio of diatom populations in freshwater is 5.5 (in Si and N weight), whereas it is about 2 for marine diatoms (Conley *et al.* 1989; Brzezinski, 1985). Whereas phytoplankton bloom intensity and duration is mainly driven by the nutrient loading, the composition of the phytoplankton blooms depends more on the nutrient ratios (Billen & Garnier, 2007). In severely human-impacted systems, N and P are in excess compared to the Si resulting almost solely from rock weathering. This imbalanced nutrient enrichment limits the growth of spring diatoms, which are then supplemented by non-siliceous algae sustained by the excess N and P. These are often undesirable, such as the mucilaginous *Phaeocystis* colonies in the southern North Sea, or toxic, such as *Dinophysis* at the coastal zone (Rousseau *et al.* 2002; Flynn, 2002; Cugier *et al.* 2005).

7.2 THE ECOLOGICAL FUNCTIONING OF THE AQUATIC CONTINUUM: MODEL CHAIN

In agreement with Vannote *et al.*'s RCC (1980), the proposed chain of models (coupled watershed and coastal models) describing the ecological functioning of the 3S and adjacent coastal zone system allows one to quantify both the nutrient fluxes and stocks along the continuum, and the coastal phytoplankton blooming response to the exported nutrients in terms of the major groups involved in seasonal successions and eutrophication.

7.2.1 A model for understanding drainage network functioning (RIVERSTRAHLER)

The RIVERSTRAHLER model (Billen *et al.* 1994; Garnier *et al.* 1995; Garnier *et al.* 2002) describes the drainage network of any river system as a combination of basins, represented as a regular scheme of confluence of tributaries of increasing stream order, each characterised by mean morphologic properties, connected to branches, represented with a higher spatial resolution. The advantage of this representation of the drainage network is that it takes into account, with reasonable calculation time, both the processes occurring in small first orders (i.e. headwater streams) and those occurring in large tributaries. The water flows in the hydrographical network are calculated from the specific discharges generated within the watershed of the different sub-basins and branches considered. The RIVERSTRAHLER model is coupled to a GIS-interface (SENEQUE), designed to calculate the seasonal and spatial variations of water quality within drainage networks ranging in scale from 10 to >100,000 km^2 (Ruelland *et al.* 2007; Billen *et al.* 2007; Thieu *et al.* 2009).

The essence of the SENEQUE/RIVERSTRAHLER model is to couple these water flows that are routed through the defined structure of basins and branches, with a model describing biological, microbiological, and physicochemical processes that occur within the water bodies. The module representing the kinetics of the processes is known as the Rive Model. The state variables comprise nutrients, oxygen, suspended matter, dissolved and particulate nonliving organic carbon, and algal, bacterial and zooplankton biomasses. Most of the processes that are important in the transformation, elimination and/or immobilisation of nutrients during their transfer within the network of rivers and streams are explicitly calculated, including algal primary production, aerobic and anaerobic organic matter degradation by planktonic as well as benthic bacteria with coupled oxidant consumption and nutrient remineralisation, nitrification and denitrification, and phosphate reversible adsorption onto suspended matter and subsequent sedimentation. Garnier *et al.* (2002) provides a detailed description of the Rive Model and of the physiological parameters used, whereas the module of the benthic processes was revisited by Thouvenot *et al.* (2007). Besides morphological and climatic constraints, the RIVERSTRAHLER model takes into account diffuse and point sources of nutrients from land-based anthropogenic sources. Nitrogen originates to a large extent from diffuse agricultural sources resulting from excess organic or mineral fertilisation of croplands, which has increased in the last 40 years (De Vries *et al.* 2011). On the other hand, phosphorus mostly originated from point sources linked to urban and industrial wastewater release, and the progress made in controlling these sources resulted in a considerable reduction of riverine phosphorus fluxes, with diffuse sources from agricultural soil now playing a significant role (Némery & Garnier 2007; Grizzetti *et al.* 2012).

7.2.2 The MIRO model for analysing ecological impacts at the coastal zone

MIRO is a mechanistic biogeochemical model describing N, P and Si cycling through aggregated components of the planktonic and benthic realms of coastal ecosystems dominated by *Phaeocystis* (Lancelot *et al.* 2005). Its structure assembles four modules describing the dynamics of phytoplankton (diatoms, nanoflagellates and *Phaeocystis*), zooplankton (copepods and micro-zooplankton), dissolved and particulate organic matter (each with two classes of biodegradability) degradation and nutrients (NO_3^-, NH_4^+, PO_4^{3-} and $Si(OH)_4$) regeneration by bacteria in the water column and the sediment. Equations and parameters were formulated based on current knowledge on the kinetics and the factors controlling the main auto- and heterotrophic processes involved in the functioning of the coastal marine ecosystem (Lancelot *et al.* 2005). The MIRO model is implemented in a multi-box framework delineated on the basis of the hydrological regime and river inputs. In order to take into account the cumulated nutrient enrichment of Atlantic waters by the Seine, Somme and Scheldt rivers, successive boxes, assumed to be homogeneous, have been chosen from the Seine Bight to the BCZ (Figure 7.1). Nutrient loads are provided by its coupling with the SENEQUE-RIVERSTRAHLER model (SR-MIRO model; Figure 7.2). This coupling is detailed in Lancelot *et al.* (2007).

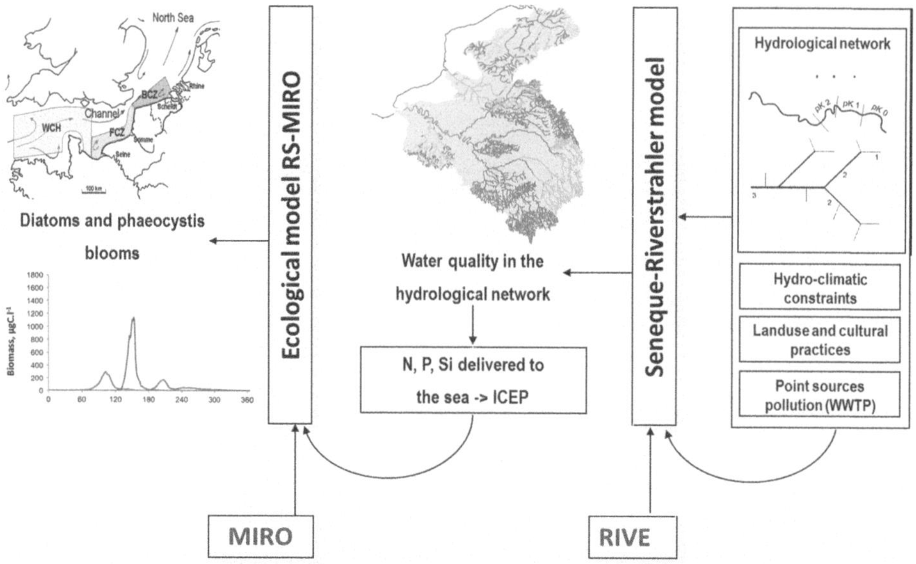

Figure 7.2 Schematic representation of the coupled models of the 3S and adjacent coastal zone (MIRO and SENEQUE/RIVERSTRAHLER. RIVE and MIRO describe the ecological processes in fresh- and marine waters, respectively.

7.3 RECONSTRUCTION OF COASTAL EUTROPHICATION IN COMPLIANCE WITH THE OSPAR CONVENTION

The OSPAR (Oslo-Paris) Convention for the Protection of the Marine Environment of the North-East Atlantic was signed in 1992 and ratified by 15 contracting parties and the European Commission (http://www.ospar.org). An agreement was adopted to take measures for reducing 1985 land-based nutrients by 50% (N and P) by 1995. In order to appraise the extent to which this target has been achieved not only in 1995, but also today, we performed a year-to-year reconstruction of the changes in hydrology, waste water treatment and agricultural practices since 1984. Chaining the SENEQUE/RIVERSTRAHLER model implemented on the Seine, Somme and Scheldt with the MIRO model describing the adjacent coastal zones (Lancelot *et al.* 2005) allowed us to quantify the link between the changes in urban and agricultural waste water management practices and algal development in the adjacent coastal zone (Passy *et al.* submitted). The changes in nutrient stoichiometry of the river inputs and the resulting shifts in nutrient limitation and *Phaeocystis* magnitude at the coastal zone were thus analysed over this 24-year period (1984–2007).

7.3.1 Nutrient input to the rivers: point and diffuse sources

7.3.1.1 Point sources of nutrients from cities and industries

Point sources are industrial and urban wastewaters discharged into the river network via wastewater treatment plants (WWTP). Each WWTP of the 3S basins (Figure 7.3) is characterised by a number of equivalent treated inhabitants and the type of treatment applied (namely primary, secondary or tertiary treatment, including P removal, nitrification and denitrification). Currently, these data were obtained from the Water Agencies in France and from the Environmental Agency of Walloon and Flemish regions in Belgium. For the 1980s and 1990s however, a reconstruction was carried out based on the population changes and the rate of connexion to sewer systems within each watershed, the year of each WWTP implementation and various assumptions (see below) on wastewater treatment progression.

Within the Scheldt and the Somme watersheds, industrial releases are assumed to be mainly collected by urban sewers while for the Seine watershed, industrial discharges are treated separately.

Severe oxygen deficits were observed in the middle of the 1980s in the lower Seine downstream from Paris (Garnier *et al.* 2001; Billen *et al.* 2007b) and in the Zenne downstream from Brussels (Billen *et al.* 1985; Garnier *et al.* 2012), resulting from carbon mineralisation by heterotrophic bacteria, but also from nitrification of the ammonium discharged with domestic waste waters, either incompletely treated in the case of the Seine or untreated for the Zenne. Regarding P, the amount discharged per inhabitant (inhabitant equivalent) was at a maximum of 4 $gP \cdot day^{-1} \cdot inh \cdot equiv^{-1}$ due to the use of polyphosphates in washing powders. Phosphorus, together with excess N, led to such high levels of algal biomass that algae paradoxically could consume more oxygen by respiration than produced by photosynthesis (Garnier & Billen, 2007).

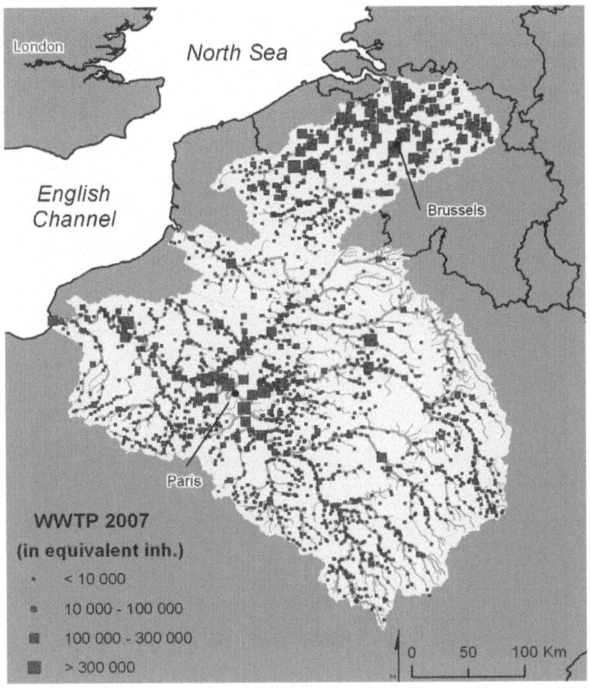

Figure 7.3 Map of waste water treatment plant distribution in the 3S basins for the year 2007. The size of the symbols indiquate the numbers in equivalent inhabitants of the various wastewater treatment plants (WWTP)

In the early 1990s, polyphosphates were banned from washing powders in Belgium (Billen *et al.* 1999); the same occurred a little later in France due to the resistance of the French industrial lobbies. In addition, a gradual P treatment in WWTP was implemented after the release of the European Directive 91/271/CEE concerning urban wastewater treatment. The specific P discharge decreased to 2 $gP \cdot day^{-1} \cdot inh \cdot equiv^{-1}$ (Garnier *et al.* 2006) and is even lower today, close to value of physiological release, that is 0.8 $gP \cdot day^{-1} \cdot capita^{-1}$ (Verbanck, 1999).

Further, recommendations to implement a tertiary treatment in order to decrease the P discharge by about 85% and the N discharge by about 70% were given to cities located within nitrate-vulnerable zones (WFD, European Water Framework Directive 2000/60/EC).

In the Seine river basin, most of Paris's waste waters are treated at the Achères station, about 50 km downstream from Paris. This WWTP, one of the largest in Europe with a treated volume of 1,700,000 m^3 of wastewater per day, was equipped with P tertiary treatment in 2000; it was upgraded to nitrify all NH_4 beginning in 2007 and denitrification up to 70% is planned to be effective in 2015.

Within the Scheldt watershed, until the beginning of this 21st century Brussels wastewaters were not treated at all and were directly released into the Zenne River (Garnier *et al.* 2012). In 2000 a first Brussels South WWTP, with a rather basic activated sludge process, was implemented a few kilometres upstream from Brussels and was fully operative in 2001. A second one, Brussels North, treating both P and N in compliance with the WFD, was implemented a few kilometres downstream from Brussels in 2007.

7.3.1.2 Diffuse sources of nutrient from agriculture

Diffuse sources are determined according to land use (based on Corine Land Cover, 1990, 2000, 2006) and agricultural practices (based on official French and Belgian agricultural statistics (www.agreste.agriculture.gouv.fr/; statbel.fgov.be). Thirteen major agricultural regions can be distinguished within the Seine, Somme and Scheldt watersheds: these are the Belgian agricultural zones (Campine, sandy region, sand-loamy region, loamy region) for the Scheldt, and a pool of French Small Agricultural Regions as defined by Mignolet *et al.* (2007) for the Seine and Somme watersheds (Figure 7.4).

For each land use type and each of the 13 agricultural regions of the drainage basins considered, the mean constant concentrations are provided for sub-root water and groundwater, respectively (in terms of N, P and Si and all other variables), which are derived from agricultural statistics, as explained below.

Changes in land use between 1990 and 2006, as described by the Corine Land Use database, are minor. The most important features are a 1.8% increase in urban areas within the Scheldt basin and a decrease in areas devoted to grasslands by 0.60%, 0.43% and 0.42% for the Seine, Somme and Scheldt basins, respectively. Other land use classes are quite stable spatially (Passy *et al.* submitted). Changes in agricultural practices were more significant.

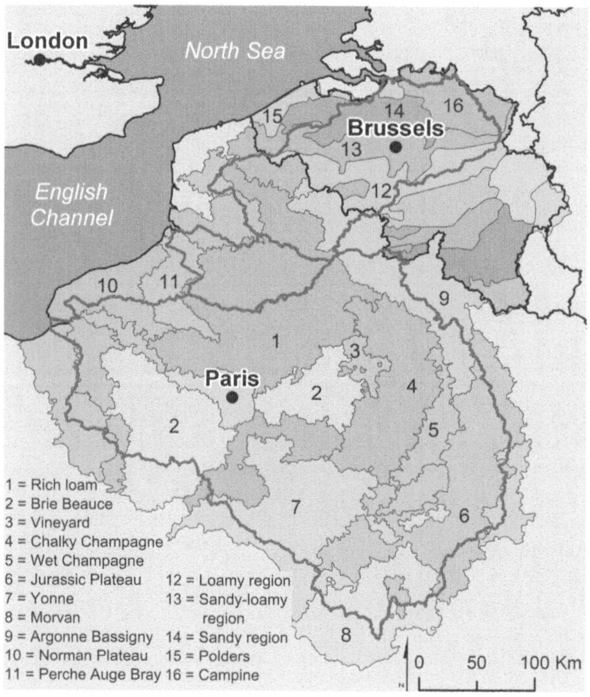

Figure 7.4 Agricultural areas within the Seine. Somme and Scheldt watersheds as taken into account in the modelling approach (cf. Passy *et al.* submitted).

The basic foundation of our approach is the existence, revealed by our very last studies, of a robust relationship, under given pedoclimatic conditions, between crop production and total N fertilisation (both integrated over the whole crop succession cycle and expressed in terms of N per agricultural area) (Passy *et al.* submitted). In well-drained arable soils, leaching is by far the main fate of the above-mentioned agricultural surplus, that is total N fertilisation – crop production; when divided by the runoff depth, it provides a good estimate of nitrate concentration of sub-root water. However, this approach cannot be used for either hydromorphic soils, like those in northern Belgium, where denitrification into inert N_2 gas is the major fate of the N surplus, or for permanent grassland soils, which are able to accumulate very large amounts of N in the soil organic pool and are subject to rather limited N leaching compared to arable land.

Based on this information, the concentration of surface runoff and baseflow produced in the different land use classes and agricultural areas of the three basins were assessed. The base flow concentration was estimated from empirical data on groundwater concentration available for the major aquifers of our study area (Figure 7.5); in regions without a large aquifer, the same concentration was taken for surface and base flow components. In the SENEQUE/RIVERSTRAHLER model, these values are used to define the diffuse sources of nutrients.

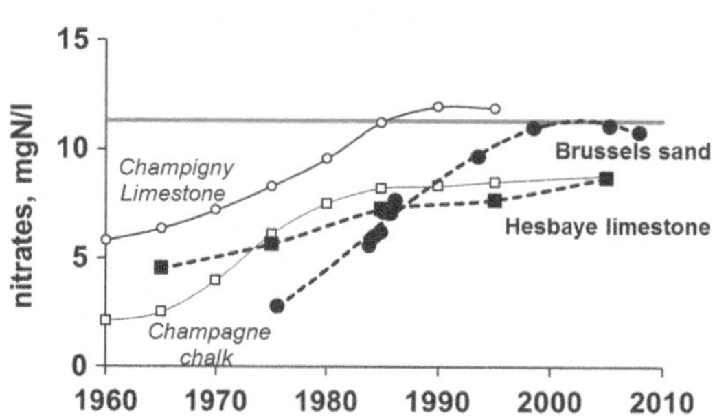

Figure 7.5 Nitrate concentrations in the major aquifers of the Seine (left, open symbols), Scheldt basins (right, dark symbols).

7.3.1.3 Modelling changes in nutrient deliveries and the effect on the southern North Sea

For the whole period considered, the SENEQUE/RIVERSTRAHLER model was run on the 3S basins using the constraints of hydroclimatic forcings, geo-morphology, the point and diffused sources of nutrients, and as described above for the 1984–2007 period (Passy *et al.* submitted), a period for which we investigated the relation between coastal eutrophication (with *Phaeocystis* blooms as a target) and human activity on the watershed. For this type of application, nutrient exports at the mouth of the 3S rivers, as calculated by the SENEQUE/RIVERSTRAHLER model, was used as input for the MIRO marine ecological model. Impacts in terms of eutrophication problems can therefore be well quantified. We will only discuss here the years 1985 and 2006, at the extremity of the studied period.

Nutrient fluxes at the river outlet

The N, P and Si export fluxes were calculated with the SENEQUE/RIVERSTRAHLER model at Poses, Abbeville and Kruibeke for the outlet of the Seine, Somme and Scheldt rivers, respectively, corresponding to the limit of tidal propagation. Comparing the values in 1984 and 2006 shows that the measures taken for phosphorus in waste water treatment plants have reduced the fluxes by more than half, so that the objectives of the OSPAR convention have been achieved (in 1984, 221, 82, 234 kgP $\cdot$ km^{-2} for the Seine, Somme and Scheldt, respectively, versus 82, 45, 118 kgP $\cdot$ km^{-2} in 2006, Table 7. 2; Figure 7.6).

Table 7.2 Nutrient fluxes (N, P, Si) at the outlet of the 3S basins. Annual mean for every 5 years during the 1985–2005 period.

Flux kg km^{-2} yr^{-1}	Seine	Somme	Scheldt
Nitrogen, N			
1985	1299	1544	1979
1990	1232	751	1652
1995	2341	1816	2504
2000	2025	2107	2445
2005	1238	1069	1450
Phosphorus, P			
1985	246	84	231
1990	137	41	129
1995	151	59	170
2000	145	67	158
2005	72	36	109
Silica, Si			
1985	577	757	716
1990	430	364	492
1995	1165	751	1052
2000	1157	735	999
2005	420	385	569

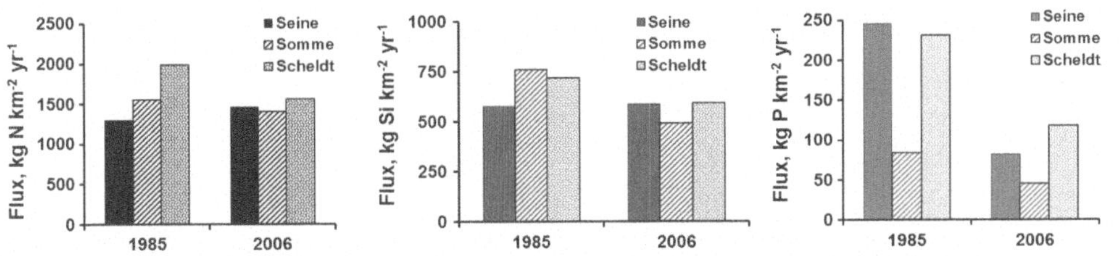

Figure 7.6 Comparison from left to right, of N, Si and P fluxes in 1985 and 2006 for the three rivers.

However, because N fluxes delivered to the coastal zone come essentially from agriculture diffuse sources, the WWTPs treating nitrogen cannot efficiently reduce the N load. Despite some reduction in fertiliser application per cropping surface area, prompted by national and European directives, the N sources from agriculture did not significantly decrease. Indeed the agricultural surplus remained very high in many regions and N concentration in large aquifers remained at high levels because of the long residence time of groundwater (30–50 years). As a result, N fluxes in 2006 and 1985 do not show any change other than that due to hydrology (1458 kgN $\cdot$ km^{-2} yr^{-1} and 1299 kgN $\cdot$ km^{-2} yr^{-1} for the Seine and 1557 kgN $\cdot$ km^{-2} yr^{-1} and 1979 kgN $\cdot$ km^{-2} yr^{-1} for the Scheldt, respectively).

The interannual variability of Si fluxes, only slightly affected by human activities, is also mainly driven by hydrological conditions. The same differences as for N are observed between the years 2006 and 1985 (587 kgSi $\cdot$ km^{-2} yr^{-1} and 577 kgSi $\cdot$ km^{-2} yr^{-1} for the Seine and 591kgSi $\cdot$ km^{-2} yr^{-1} and 716 kgSi $\cdot$ km^{-2} yr^{-1} for the Scheldt, respectively).

The spatial distribution of modelled mean P concentration estimated for the 3S basin in 1985 and 2006 is shown in Figure 7.7. Mean concentrations are shown according to the quality colour code (SEQ-Eau, 2003).

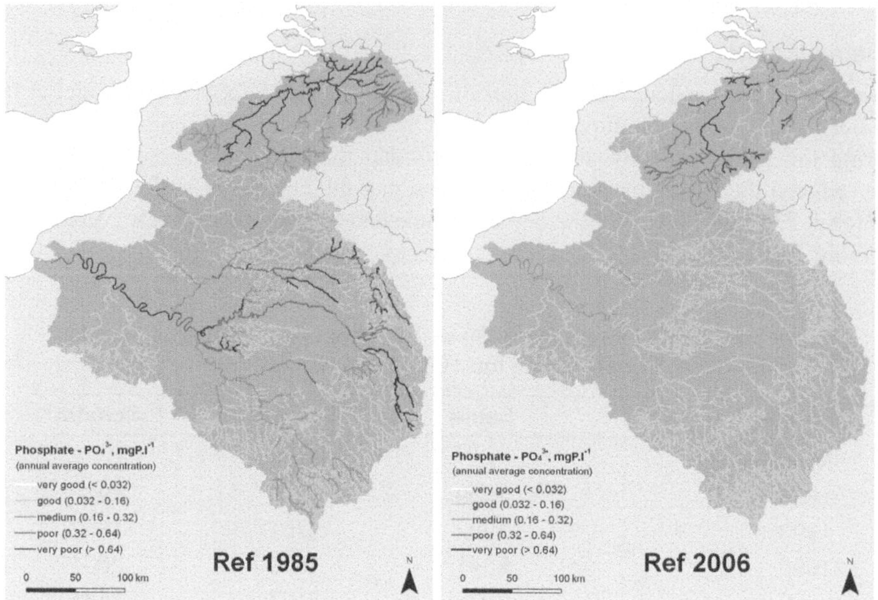

Figure 7.7 Comparison of phosphorus concentrations in the drainage network of the 3S basins for 1985 and 2006.

A clear net improvement is simulated in 2006, mainly in the lower Seine River downstream from Paris. A slight improvement in the lower Somme River is observed. Concerning the Scheldt River, a significant change is suggested, with concentrations greater than 0.64 mgP L^{-1} in most tributaries in 1985 reduced to 0.032–0.16 mgP L^{-1} in 2006, except in some sectors, such as the Zenne River impacted by the city of Brussels (Figure 7.7).

Phytoplankton changes at the coastal zone

When combined with SENEQUE/RIVERSTRAHLER, the MIRO model shows the expected phytoplankton spring succession characterised by a first Si-limited diatom bloom supplemented by a higher bloom of *Phaeocystis* colonies whose growth is controlled by excess nitrates (Rousseau *et al.* 2002). Changes in nutrients between the two extreme dates of the 24-year period show a significant decrease in the *Phaeocystis* blooms and of summer diatoms to a less extend (Figure 7.8).

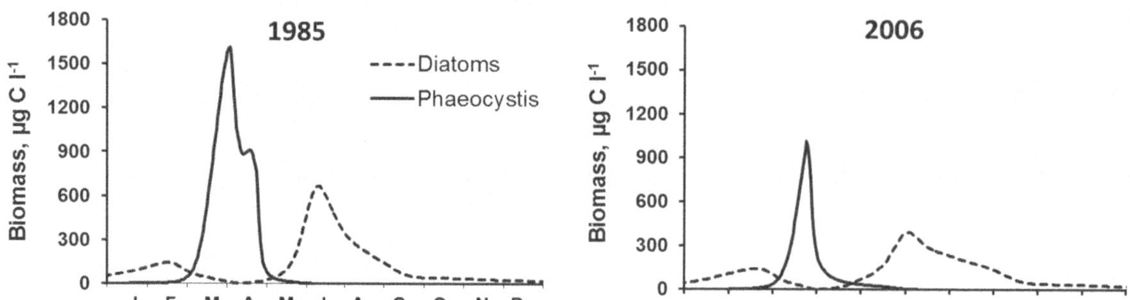

Figure 7.8 Simulation of the seasonal successions of diatoms and non-diatoms at the southern North Sea in 1985 and 2006.

As suggested by Lancelot *et al.* (2009), the maximum abundance of 4 10^6 cells L^{-1} describing a healthy *Phaeocystis* ecosystem can be used for scaling the undesirable effect of *Phaeocystis* blooms both as maximum abundance reached and as duration, that is the number of days with *Phaeocystis* cells higher than the indicator (Figure 7.9). Regarding the maximum *Phaeocystis* abundance, a decrease surpasses 30%, that is 56.5 vs. 36.8 10^6 cells L^{-1} in 1985 and 2006, respectively, a figure still well above the reference of 4.10^6 cells L^{-1}, however, found to be a reference characterising the lack of eutrophication (Lancelot *et al.* 2009). The number of days with foam on the beach is reduced in a similar proportion, to 30 days in 2006 against 53 days in 1984 (Figure 7.9).

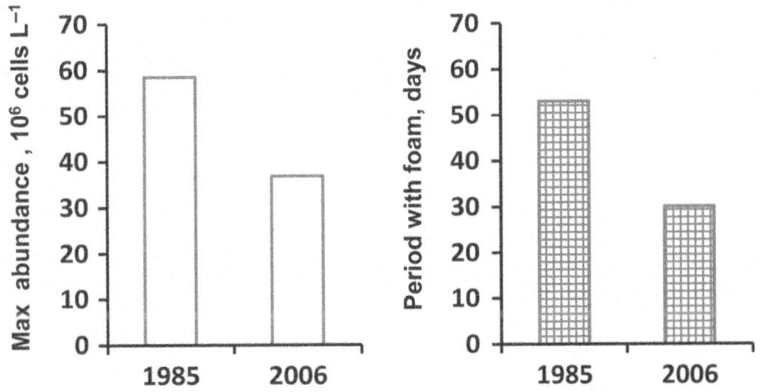

Figure 7.9 Maximum abundance of Phaeocystis in 1985 and 2006 (left). Number of days with the presence of foam in 1985 and 2006 (right).

7.4 WHAT CAN BE DONE TO IMPROVE THE SITUATION?

Early in the AWARE process, citizens understood the scope and stakes of the integrated model and were able to discuss the results of this approach and to appreciate the changes in water quality in relation to the measures taken throughout the period.

From exhaustive discussions with citizens, stakeholders and scientists during the AWARE conference and the local workshops, nutrient reduction options – *beyond those already prescribed by national and European directives* – have emerged for the upstream watersheds in order to further the reduction of nutrients in the coastal zone, taking into account the issues raised by the "Grenelle de l'Environnement" in France and on-going experiments on the watersheds that include protected areas devoted to drinking water production. For these watersheds, it was decided to banish mineral fertilisation of agriculture for organic agricultural practices.

It was decided to explore and compare two realistic scenarios with the year 2006 taken as a reference year.

7.4.1 Improving waste water treatment

Citizens suggested first exploring a scenario simulating the implementation of the European Directive dealing with the improvement of wastewater treatment plants (91/271/CEE). This directive stipulates that N and P be decreased by 70 and 90% within each WWTP treating more than 2000 inhabitant equivalents. This scenario allowed citizens to appraise the achievements of this measure as well as the respective weight of point and diffuse pollution in terms of N and P. The SENEQUE/RIVERSTRAHLER – MIRO modelling chain was run taking into account these new treatments over the three watersheds. Nitrate concentrations within the river network and the intensity and length of blooms in the coastal zone are shown in Figure 7.10.

As explained above, P for the most part originates from point sources as urban effluents, while N originates essentially from diffuse sources. Consequently, the retrofitting of WWTP only leads to a minor decrease of N within the hydrological network. Water quality is only improved in a few tributaries of the Somme and Scheldt basins. Yet some improvements can be observed in the coastal zone. Although the magnitude of the bloom shows little decrease with this scenario, the duration of the bloom shifts from 30 days in the reference situation to 28 days. Finally this scenario shows that no significant N decrease will be obtained by improving the quality of urban or industrial effluents. Therefore, other measures dealing with the diffuse pollution originating from nutrients losses from agriculture have to be considered.

7.4.2 Reducing nutrient losses from agriculture

The challenge has thus moved from wastewater management to agricultural practices. In 2007, France adopted the 'Grenelle de l"Environnement", a corpus of laws legislating key environmental issues such as the quality of continental and coastal waters. One of the targets was to promote agriculture reducing the use of synthetic fertilisers and pesticides. Organic farming, in particular, not only forbids any use of pesticides, but also releases much less N in leaching water when properly implemented (Stanhill 1990; Drinkwater *et al.* 2000; Hansen *et al.* 2000; Korsaeth *et al.* 2000). More precisely, the "Grenelle" strongly recommended organic farming within catchments dedicated as drinking water protection areas (DWPAs), as implemented for several years by the city of Munich (http://www.partagedeseaux.info/article48.html). These areas correspond to the watershed located upstream from wells supplying the population in fresh water. DWPAs are quite difficult to delineate due to lack of hydrogeological knowledge in some parts of the basins, but especially because of certain political, economic and societal issues. Indeed, the promotion of an alternative agriculture could spark resistance from some stakeholders. Despite these uncertainties on the exact limits of DWPAs, a scenario implementing organic farming on DWPAs was constructed (Figure 7.11).

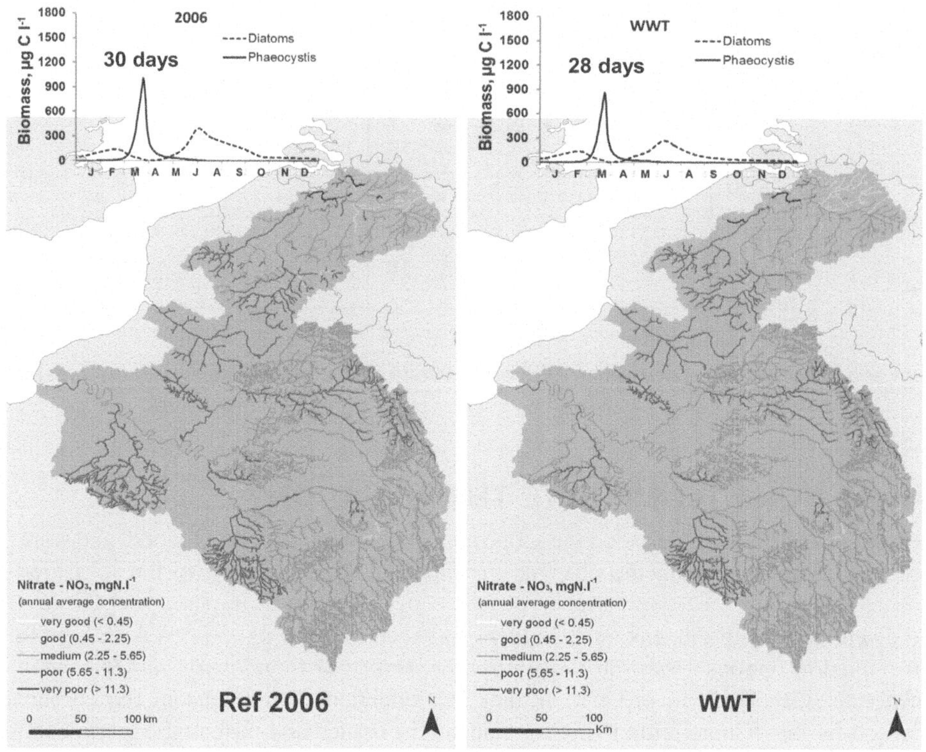

Figure 7.10 Comparison in nitrate concentrations in the drainage network of the 3S basins for the year 2006 and for a scenario with the implementation of the EU urban waste water directive (WWT) as it should be in 2015. The seasonal evolution of diatoms and Phaeocystis is indicated as well as the duration of the undesirable bloom and hence the foam risk.

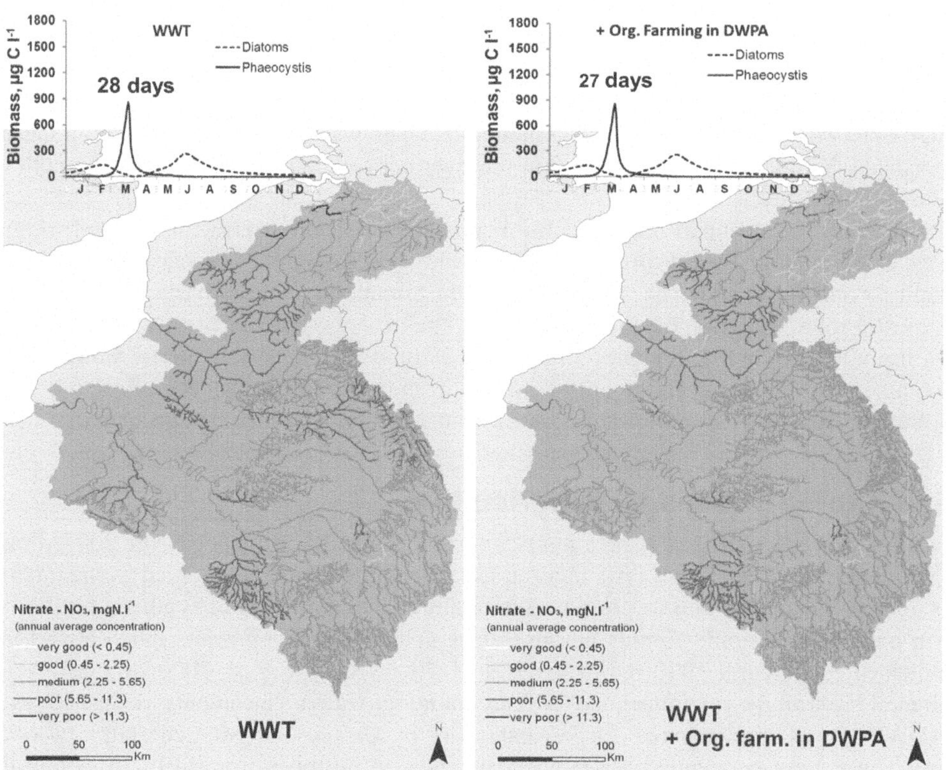

Figure 7.11 Nitrate concentrations in the drainage network of the 3S basins comparing two scenarios, i) implementation of the EU urban waste water directive as it should be in 2015 (left), implementation of N efficient organic farming (releasing sub-root water at 3 mgN-NO_3 L^{-1}) in the watersheds devoted to drinking water protection areas. The seasonal evolution of diatoms and Phaeocystis is indicated as well as the duration of the undesirable bloom and hence the foam risk.

Figure 7.11 compares the water quality, in terms of nitrate concentrations, that will be reached when all WWTPs achieve the European norm (left) with that possibly obtained by implementing N-efficient organic farming. As this scenario includes agricultural sources and therefore diffuse pollution, most impacts can be observed on nitrate concentrations. Within the river networks, this measure could reduce the nitrate concentrations in several tributaries in the upper Oise and Eure basins, as well as in some rivers of the Somme and Scheldt watersheds.

Impacts on coastal zones are even less significant. The magnitude of *Phaeocystis* bloom hardly decreases and the *Phaeocystis* bloom duration remains unchanged at 27 days. Unfortunately, the implementation of N-efficient organic farming in DWPAs, although it would provide significant improvements in freshwater quality, seems to be insufficient to reduce eutrophication of the Belgian – French coastal zone.

7.4.3. Further changes in the agricultural sector: no or low fertilisation, drainage ponds

Changing agricultural practices, which consists in applying less fertiliser, thus leading to less nitrate leaching, is a preventive approach. This kind of approach can be combined with curative solutions. One possible way to decrease nitrate concentrations within the hydrological network is to implement stagnant water bodies (ponds and reservoirs) in the river network or at the outlet of agricultural drains. Indeed the nitrate retention and elimination capacity of ponds, lakes and reservoirs has been common knowledge for many years (Dillon & Molot, 1990; Garnier *et al.* 1999). A large part of nitrate routing by these water bodies can be eliminated by denitrification.

Citizens have shown interest in exploring this kind of scenario. The question was, however, where to locate such ponds in the basins. As the introduction of ponds on every river is unrealistic, the scenario was constructed with the history of the watersheds in mind. Examination of maps from the 18th century, in particular Cassini's map covering France in its entirety, pointed out the presence of more than 2500 ponds at the end of the 18th century. These ponds were located on small rivers upstream in the watershed in regions with impermeable geological substrate. They were used for mills, fishing or timber rafting. They were filled during the 19th century and most of them have disappeared today. Based on this historical analysis, a pond reintroduction measure was elaborated and tested on the 3S watersheds (Passy *et al.* 2012), allowing us to run the scenario of coupling organic farming on DWPAs, the European Urban Waste water Directive (WWT) and pond reintroduction (Figure 7.12 left).

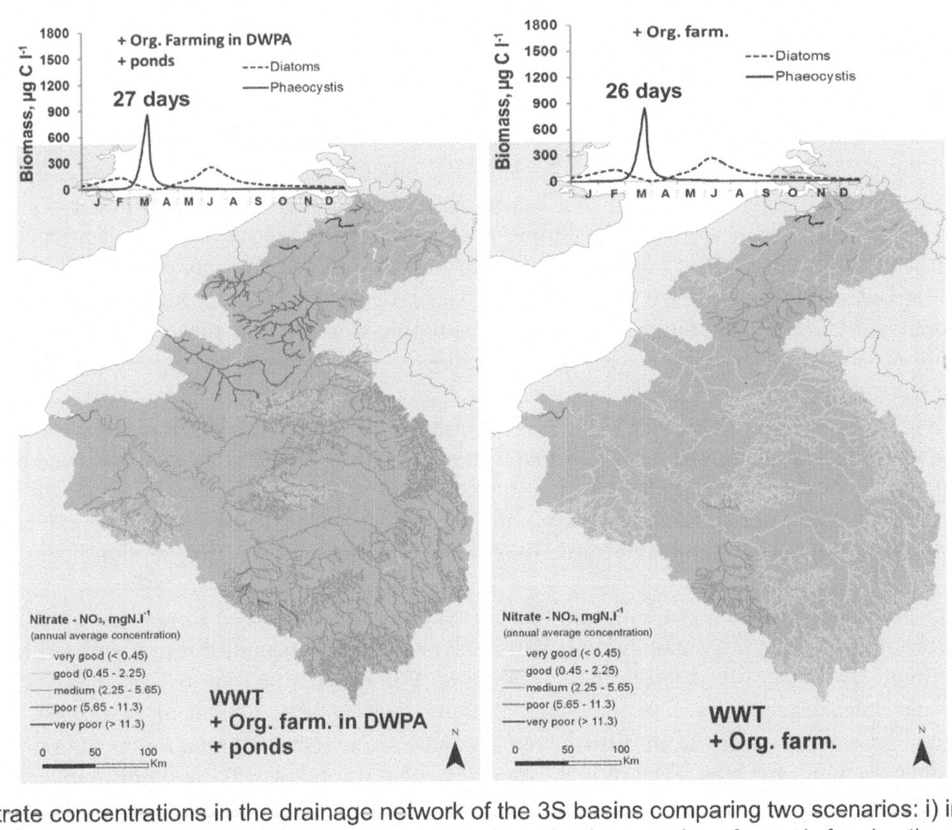

Figure 7.12 Nitrate concentrations in the drainage network of the 3S basins comparing two scenarios: i) implementation of organic farming on DWPAs, WWT, and pond reintroduction (left) and implementation of organic farming throughout the Seine, Somme and Scheldt watersheds. The seasonal evolution of diatoms and Phaeocystis is indicated as well as the duration of the undesirable bloom and hence the foam risk.

This scenario clearly improves the state of nitrate contamination, particularly in some upper parts of the Seine basin, where rivers are rated "good" (see Figure 7.11 for comparison). The situation is also improved in the Scheldt watershed where the hydrological network shifts from 'poor' to 'medium' and even to 'good' in the eastern part of the basin. The impact at the coast is disappointing, however, as the maximum blooms and their duration remained unchanged.

As seen above, the challenge of managing diffuse sources requires strong motivation. To form an idea of the best possible achievable result, citizens asked for an idealised scenario implementing organic farming on all arable lands of the three watersheds. This scenario had already been explored by Thieu *et al.* (2011), and used by Lancelot *et al.* (2011), examining its impact at the coast. The implementation of organic farming throughout the 3S watersheds is not realistic due to its many political, economic and societal obstructions, but this kind of theoretical scenario raises awareness of the best possible results (Figure 7.12, right).

As shown in Figure 7.12, nitrate concentrations decrease spectacularly. The whole upper Seine basin shifts from "medium" or "poor" to "good". The lower part of the Seine becomes "good". The whole Somme watershed is 'good' and the main part of the Scheldt basin becomes "medium" or "good". The situation remains less satisfactory within the Scheldt basin due to this region's more intensive farming and to the basin's higher population density. Despite these positive results on hydrological networks, the intensity of *Phaeocystis* blooms remains high. Even the bloom duration would not decrease much, shifting only from 27 days to 26 days.

7.5 SUSTAINABILITY OF THE MEASURES TO BE TAKEN

While coastal eutrophication is an obvious environmental problem in itself, compromising local economic activities, the increase of nutrient (N and P) contamination and the consequent reduction in river and groundwater quality in the watersheds is also a serious environmental threat for ecosystem health and for producing drinking water for the inhabitants.

Algal blooms in large rivers may still cause severe water quality problems and problems in drinking water production. Nitrate contamination of groundwater resources is also becoming alarming because it often approaches the drinking water standard of 50 $mgNO_3\ L^{-1}$ (=11 $mgN\ L^{-1}$).

During the course of the AWARE project, citizens, scientists, policy makers and other stakeholders concerned by human-caused deterioration of water quality actively participated in workshops held at both local and European levels.

The 3S-North Sea local workshops and conference were very successful. The local workshops were characterised by active and enthusiastic discussions among scientists, citizens and stakeholders, from which emerged the majors ideas for scenario construction, the local conference, also open to experts, policy-makers and the public, helped to disseminate the main messages from the scenarios tested. The event was even covered by several local media. The results were widely debated, in particular via the citizens' declaration.

The feasibility and outcomes of different measures to improve water quality were discussed beyond the results. Most people involved in the AWARE project, as well as the AWARE process, pointed out the importance of increasing the connectivity not only amongst water resource actors but much more broadly. How the population would be fed was clearly questioned, especially because of the pollution originating from agriculture. In particular, the excessive production and consumption of meat was pointed out, and a common agreement was met for changing behaviour as recommended by the Barsac declaration (www.nine-esf.org/barsac-declaration), promoting a 50% reduction in animal protein intake (Billen *et al.* 2011). Although nutrients and eutrophication were at the centre of the AWARE project, other related pollutions came under debate, such as pesticides, a major threat for farmers, a population category showing increasing cancer rates.

As a collaborative, solution-oriented, approach, the AWARE project has succeeded in delivering recommendations and scenarios to European and local policy-makers, and ultimately making all stakeholders aware of the need to involve citizens and the broad public in the dialog, once they have been informed of the context of the identified question. Eutrophication caused by excess nutrients was presented with easily comprehensible scientific presentations and common work on scenarios proved to be a useful communication tool. Knowledge sharing was one of the significant achievements of the project.

The two distinct pathways by which nutrients are transferred from the watershed to surface water are a true key to understanding and combatting nitrate contamination. Diffuse contamination occurs through soil leaching and erosion, thus reflecting the intensity of the use of fertiliser in agriculture. Point contamination results from the direct discharge of urban waste water into surface water. In the Seine, Somme and Scheldt (Thieu *et al.* 2009), we have clearly evidenced that nitrate essentially comes from diffuse sources and even a 100% elimination at the point sources would not solve nitrate contamination problems, even though some agricultural lobbies still deny this evidence. The EU-Urban Waste Water Directive (UWWTD) has required a 70% nitrate reduction, which is currently being applied. A further reduction is technically difficult because denitrification is a microbial process which requires organic carbon and anoxia (Knowles, 1982) for optimum conditions, to avoid greenhouse gas emissions (N_2O), an intermediate product (Tallec *et al.* 2008).

In the 3S basins, point sources of urban and industrial nutrient pollution are now treated nearly in compliance with EU directives. Agriculture, however, is still a growing source of nutrient hydrosystem contamination. In particular, industrial agriculture and its use of inorganic N fertilisers leads to a massive introduction of reactive nitrogen into the aquatic system, which subsequently cascades through a number of environmental compartments until it is partly denitrified and emitted to the atmosphere (Galloway & Cowling, 2002; Sutton *et al.* 2011).

Nitrate contamination of water resources, but also biodiversity loss, atmospheric pollution by ammonia and ammonium nitrate micro-particles, and nitrous oxide emissions are among the major threats resulting from this acceleration of the nitrogen cycle. Due to the long residence time of some of the environmental compartments (soils, vadose zones, aquifers, etc.), N equilibrium between soils and aquifers will not be reached until several decades after the onset of industrial agriculture, meaning we can still expect a sustained N contamination of the ground- and surface waters. Further, the debate must include the question of urban water supply (including drinking water) besides eutrophication, because water abstraction for urban use might well compete with that required for agriculture and food production in the same land areas, particularly in the densely populated regions such as the centre of the Parisian and the Scheldt basins. In the coming decades, in addition to searching for good-quality water for domestic use, water resources will be a big challenge due to global climate change (McDonald *et al.* 2011).

To overcome these problems in the next 50 years, a change should be initiated now in agriculture. A political consensus on this matter is very difficult to achieve (Leridon & de Marsily, 2011; Swinnen & Squicciarini, 2012). The regional or watershed scale is probably the most suitable scale to resolve such questions, because at this level it is possible to achieve coherence in the decision to implement some measure as well as the management of the chosen measure. The regional scale, however, is itself both nested in several higher scales (national, continental, global) and relies on small-scale decisions (farmers, citizens, etc.), (Wilson, 2009). Directly involving citizens in the discussion of these issues at the regional scale, organising a dialog between citizens, policy-makers, stakeholders and scientists proved not only possible but also quite fruitful.

REFERENCES

Billen G., Somville M., De Becker E. and Servais P. (1985). A nitrogen budget of the Scheldt hydrographical basin. *Netherlands Journal of Sea Research*, **19**, 223–230.

Billen G., Garnier J. and Hanset Ph. (1994). Modelling phytoplankton development in whole drainage networks: the RIVERSTRAHLER model applied to the Seine river system. *Hydrobiologia*, **289**, 119–137.

Billen G. and Garnier J. (1999). Nitrogen transfers through the Seine drainage network: a budget based on the application of the Riverstrahler model. *Hydrobiologia*, **410**, 139–150.

Billen G., Garnier J., Ficht A. and Cun C. (2001). Modelling the response of water quality in the Seine Estuary to human activity in its watershed over the last 50 years. *Estuaries*, **24**(6), 977–993.

Billen G., Garnier J. and Rousseau V. (2005). Nutrient fluxes and water quality in the drainage network of the Scheldt basin over the last 50 years. *Hydrobiologia*, **540**, 47–67.

Billen G. and Garnier J. (2007). River basin nutrient delivery to the coastal sea: assessing its potential to sustain new production of non siliceous algae. *Marine Chemistry*, **106**, 148–160. doi: 10.1016/j.marchem.2006.12.017.

Billen G., Garnier J., Némery J., Sebilo M., Sferratore A., Benoit P., Barles S. and Benoit M. (2007). A long term view of nutrient transfers through the Seine river continuum. *Science of the Total Environment*, **275**, 80–97.

Brzezinski M. A. (1985). The Si:C:N ratio of marine diatoms: interspecific variability and the effect of some environmental variables. *Journal of Phycology*, **21**, 347–357.

Cadée G. C. and Hegeman J. (1991). Historical phytoplankton data of the Marsdiep. *Hydrobiological Bulletin*, **24**, 111–118.

Cloern J. E. (2001). Our evolving conceptual model of the coastal eutrophication problem. *Marine Ecology Progress Series*, **210**, 223–253.

Conley D. J., Kilham S. S. and Theriot E. C. (1989). Differences in silica content between marine and freshwater diatoms. *Limnology and Oceanography*, **34**, 205–213.

Cowell S. J. and Parkinson S. (2003). Localisation of UK food production: an analysis using land area and energy as indicators. *Agriculture, Ecosystems and Environment*, **94**, 221–236.

Cugier Ph., Billen G., Guillaud J. F., Garnier J. and Ménesguen A. (2005). Modelling the eutrophication of the Seine Bight (France) under historical, present and future riverine nutrient loading. *Journal of Hydrology*, **304**, 381–396.

Diaz R. J. (2001). Overview of hypoxia around the world. *Journal of Environmental Quality*, **30**, 275–281.

Dillon P. and Molot L. (1990). The role of ammonium and nitrate retention in the acidification of lakes and forested catchments. *Biogeochemistry*, **11**, 23–43.

Drinkwater L. E., Wagoner P. and Sarrantonio M. (1998). Legume-based cropping systems have reduced carbon and nitrogen losses. *Nature*, **396**, 262–265.

Even S., Billen G., Bacq N., Théry S., Ruelland D., Garnier J., Cugier Ph., Poulin M., Blanc S., Lamy F. and Paffoni C. (2007). New modelling tools for water quality management of regional river systems: simulating a prospective scenario in the scope of the European Water Framework Directive in the Seine River Basin. *The Science of the Total Environment*, **375**, 274–291.

Flynn K. J. (2002). Modeling marine phytoplankton growth under eutrophic conditions. *Journal of Sea Research*, **54**, 92–103.

Galloway J. N. and Cowling E. B. (2002). Reactive nitrogen and the world: 200 years of change. *Ambio*, **31**, 64–71.

Garnier J., Billen G. and Coste M. (1995). Seasonnal succession of diatoms and Chlorophyceae in the drainage network of the river Seine: observations and modelling. *Limnology and Oceanography*, **40**, 750–765.

Garnier J., Leporcq B., Sanchez N. and Philippon X. (1999). Biogeochemical mass-balances (C, N, P, Si) in three large reservoirs of the Seine basin (France). *Biogeochemistry*, **47**, 119–146.

Garnier J., Servais P., Billen G., Akopian M. and Brion N. (2001). The oxygen budget in the Seine estuary: balance between photosynthesis and degradation of organic matter. *Estuaries*, **24**(6), 964–977.

Garnier J., Billen G., Hannon E., Fonbonne S., Videnina Y. and Soulie M. (2002). Modeling transfer and retention of nutrients in the drainage network of the Danube River. *Estuarine, Coastal and Shelf Science*, **54**, 285–308.

Garnier J., Laroche L. and Pinault S. (2006). Determining the domestic specific loads of two wastewater plants of the Paris conurbation (France) with contrasted treatments: a step for exploring the effects of the application of the European Directive. *Water Research*, **40**, 3257–3266.

Garnier J. and Billen G. (2007). Production vs. Respiration in river systems: an indicator of a 'good ecological status' evaluation. *The Science of the Total Environment*, **375**, 110–124.

Garnier J., Brion N., Callens J., Passy P., Deligne C., Billen G., Servais P. and Billen C. (2012). Modelling historical changes in nutrient delivery and water quality of the Zenne river (1790s–2010): the role of land use, waterscape and urban wastewater management. *Journal of Marine Systems*. doi: 10.1016/j.jmarsys.2012.04.001 (in press).

Glibert P. M., Mayorga E. and Seitzinger S. P. (2008). Prorocentrum minimum tracks anthropogenic nitrogen and phosphorus inputs on a global basis: application of spatially explicit nutrient export models. *Harmful Algae*, **8**, 33–38.

Grizzetti B., Bouraoui F. and Aloe A. (2012). Changes of nitrogen and phosphorus loads to European seas. *Global Change Biology*, **18**, 769–782. doi: 10.1111/j.1365-2486.2011.02576.x.

Hansen B., Kristensen E., Grant R., Høgh-Jensen H., Simmelsgaard S. and Olesen J. (2000). Nitrogen leaching from conventional versus organic farming systems – a systems modelling approach. *European Journal of Agronomy*, **13**, 65–82.

Howarth R. W., Billen G., Swaney D., Townsend A., Jaworski N., Lajtha K., Downing J. A., Elmgren R., Caraco N., Jordan T., Berendse F., Freney J., Kudeyarov V., Murdoch P. and Zhao-Liang Z. (1996). Regional nitrogen budgets and riverine N & P fluxes for the drainages to the North Atlantic Ocean: natural and human influences. *Biogeochemistry*, **35**, 75–139.

Howarth R. W. (2008). Coastal nitrogen pollution: a review of sources and trends globally and regionally. *Harmful Algae*, **8**, 14–20.

Justic D. (1991). Hypoxic conditions in the northern Adriatic sea: historical development and ecological significance. In: Modern and Ancient Continental Shelf Anoxia, R. V. Tyson and T. H. Pearson (eds.), Geological Society, London, pp. 95–105, Special Publication No. 58.

Justic D., Rabalais N. N., Turner R. E. and Dortch Q. (1995a). Changes in nutrient structure of river-dominated coastal waters: stoichiometric nutrient balance and its consequences. *Estuarine, Coastal and Shelf Science*, **40**, 339–356.

Justic D., Rabalais N. N. and Turner R. E. (1995b). Stoichiometric nutrient balance and origin of coastal eutrophication. *Marine Pollution Bulletin*, **30**, 41–46.

Knowles R. (1982). Denitrification. *Microbiological Reviews*, **46**, 43–70.

Korsaeth A. and Eltun R. (2000). Nitrogen mass balances in conventional, integrated and ecological cropping systems and the relationship between balance calculations and nitrogen runoff in an 8-year field experiment in Norway. *Agriculture, Ecosystems & Environment*, **79**, 199–214.

Lancelot C., Billen G., Sournia A., Weisse T., Colijn F., Veldhuis M., Davies A. and Wassman P. (1987). *Phaeocystis* blooms and nutrient enrichment in the continental coastal zones of the North Sea. *Ambio*, **16**, 38–46.

Lancelot C., Spitz Y., Gypens N., Ruddick K., Becquevort S., Rousseau V., Lacroix G. and Billen G. (2005). Modelling diatom and *Phaeocystis* blooms and nutrient cycles in the Southern Bight of the North Sea: the MIRO model. *Marine Ecology, Progress Series*, **289**, 63–78.

Lancelot C., Gypens N., Billen G., Garnier J. and Roubeix V. (2007). Linking marine eutrophication to land use: an integrated river-ocean mathematical tool: The Southern Bight of the North Sea over the past 50 years. *Journal of Marine Systems*, **64**, 216–228. doi: 10.1016/j.jmarsys.2006.03.010.

Lancelot C., Thieu V., Polard A., Garnier J., Billen G., Hecq W. and Gypens N. (2011). Ecological and economic effectiveness of nutrient reduction policies on coastal *Phaeocystis* colony blooms in the Southern North Sea: an integrated modeling approach. *Sciences of the Total Environnement*, **409**, 2179–2191. doi: 10.1016/j.scitotenv.2011.02.023.

Leridon H. and de Marsily G. (2011). Démographie, climat et alimentation mondiale. In: Rapport sur la Science et la Technologie N°32. Académie des Sciences. EDP Sciences (ed.), Les Ulis (91), France, p. 313. ISBN: 978-2-7598-05581-5.

Marchetti R. (1991). Algal blooms and gel production in the Adriatic sea. In: Eutrophication Related Phenomena in the Adriatic Sea and in other Mediterranean Coastal Zones, H. Barth and L. Fegan (eds.), Water Pollution Research Reports Series n°16, CEC DGXII, Brussels, pp. 21–42.

Mignolet C., Schott C. and Benoît M. (2007). Spatial dynamics of farming practices in the Seine basin: methods for agronomic approaches on a regional scale. *Science of The Total Environment*, 375, 13–32. doi: 10.1016/j.scitotenv.2006.12.004.

Némery J. and Garnier J. (2007). Origin and fate of phosphorus in the Seine watershed (France): agricultural and hydrographic P budgets. *Journal of Geophysical Research*, **112**, G03012, doi: 10.1029/2006JG000331.

Officer C. B. and Ryther J. H. (1980). The possible importance of silicon in marine eutrophication. *Marine Ecology Progress Series*, **3**, 83–91.

OSPAR Commission (2000). Quality Status Report. OSPAR Commission.

Passy P., Gypens N., Billen G., Garnier J., Thieu V., Rousseau V., Callens J., Parent J.-Y. and Lancelot C. (submitted). A model reconstruction of riverine nutrient fluxes and eutrophication in the Belgian Coastal Zone since 1984. *Journal of Marine Systems*.

Passy P., Garnier J., Billen G., Fesneau C. and Tournebize J. (2012). Restoration of ponds in rural landscapes: modelling the effect on nitrate contamination of surface water (the Seine watershed, France). *Science of the Total Environment*. doi: 10.1016/j.scitotenv.2012.04.035.

Rabouille C., Conley D. J., Dai M. H., Cai W.-J., Chen C. T. A., Lansard B., Green R., Yin K., Harrison P. J., Dagg M. and McKee B. (2008). Comparison of hypoxia among four river-dominated ocean margins: the Changjiang (Yangtze), Mississippi, Pearl, and Rhône rivers. *Continental Shelf Research*, **28**, 1527–1537.

Redfield A. C., Ketchum B. H. and Richards F. A. (1963). The influence of organisms on the composition of sea-water. In: The Sea, M. N. Hill (ed.), John Wiley & Sons, New York, pp. 12–37.

Ruelland D., Billen G., Brunstein D. and Garnier J. (2007). SENEQUE 3: a GIS interface to the RIVERSTRAHLER model of the biogeochemical functioning of river systems. *The Science of the Total Environment*, **375**, 257–273.

Rousseau V., Becquevort S., Parent J.-Y., Gasparini S., Daro M.-H., Tackx M. and Lancelot C. (2000). Trophic efficiency of the planktonic food web in a coastal ecosystem dominated by *Phaeocystis* colonies. *Journal of Sea Research*, **43**, 357–372.

Rousseau V., Leynaert A., Daoud N. and Lancelot C. (2002). Diatom succession, silicification and silicic acid availability in Belgian coastal waters (Southern North Sea). *Marine Ecology Progress Series*, **236**, 61–73.

Stanhill G. (1990). The comparative productivity of organic agriculture. *Agriculture, Ecosystems & Environment*, **30**, 1–26.

Swinnen J. and Squicciarini P. (2012). Mixed messages on prices and food security. *Science*, **335**, 405–406.

Sebilo M., Billen G., Grably M. and Mariotti A. (2003). Isotopic composition of nitrate-nitrogen as a marker of riparian and benthic denitrification at the scale of the whole Seine River system. *Biogeochemistry*, **63**, 35–51.

SEQ-Eau (2003). Le Système d'Evaluation de la Qualité des cours d'eau (SEQ). Grilles d'évaluation version 2. MEDD & Agences de l'eau.

Smayda T. J. (1990). Novel and nuisance phytoplankton blooms in the sea: evidence for a global epidemic. In: Toxic Marine Phytoplankton, E. Granéli, B. Sundstrom, L. Edler and D. M. Anderson (eds.), Elsevier, New York, pp. 29–40.

Sutton M., Howard C., Erisman J. W., Billen G., Bleeker A., Grennfelt P., van Grinsven H. and Grizzetti B. (eds.) (2011). The European Nitrogen Assessment: Sources, Effects and Policy Perspectives. Cambridge University Press, Cambridge, pp. 1–612.

Tallec G., Garnier J., Billen G. and Gousailles M. (2008). Nitrous oxide emissions from denitrifying activated sludge of urban wastewater treatment plants, under anoxia and low oxygenation. *Bioresource Technology*, **99**, 2200–2209. doi: 10.1016/j.biortech.2007.05.025.

Tilman D., Fargione J., Wolff B., D'Antonio C., Dobson A., Howarth R., Schindler D., Schlesinger W. H., Simberloff D. and Swackhamer D. (2001). Forecasting agriculturally driven global environmental change. *Science*, **292**, 281–284.

Thieu V., Billen G. and Garnier J. (2009). Nutrient transfer in three contrasting NW European watersheds: the Seine, Somme, and Scheldt Rivers. A comparative application of the Seneque/Riverstrahler model. *Water Research*, **43**(6), 1740–1754.

Thouvenot M., Billen G. and Garnier J. (2007). Modelling nutrient exchange at the sediment–water interface of River Systems. *Journal of Hydrology*, **341**, 55–78.

UWWTD (Urban wastewater treatment Directive (1991). 91/271/CEE du Conseil, du 21 mai 1991, relative au traitement des eaux urbaines résiduaires. JO L 135, du 30.5.1991, pp. 40–52.

Verbanck M., Vanderborght J.-P. and Wollast R. (1994). Major ion content of urban wastewater: assessment of per capita loading. *Journal of the Water Pollution Control Federation*, **61**, 1722–1728.

Vollenweider R. A. (1968). Water management research. Scientific fundamentals of the eutrophication of lakes and flowing waters with particular reference to nitrogen and phosphorus as factors in eutrophication, Tech. Rep., DA 5/SCI/68.27, OECD, Paris, p. 250.

de Vries F. T., Van Groenigen J. W., Hoffland E. and Bloem J. (2011). Nitrogen losses from two grassland soils with different fungal biomass. *Soil Biology and Biochemistry*, **43**, 997–1005.

Wilson G. A. (2009). The spatiality of multifunctional agriculture: a human geography perspective. *Geoforum*, **40**, 269–280.

WFD (Water Framework Directive). OJ L327/1, 22.12. 2000 of the European Parliament and of the Council establishing a framework for the Community action in the field of water policy, pp. 1–72.

Chapter 8

The Sacca di Goro: a cooperative decision making experiment for a sustainable lagoon exploitation

Pierluigi Viaroli, Gianmarco Giordani, Chiara Mocenni, Emiliano Sparacino, Stefano Lovo and Silvano Bencivelli

8.1 INTRODUCTION: KEY ISSUES AND RATIONALE

In the Sacca di Goro lagoon, clam farming is the main activity which support both the local economy and social system. The management of the aquatic ecosystem and the preservation of good water quality are thus perceived as key issues by stakeholders and citizens, although their points of view can greatly differ each others. Water quality is critical for producing safe seafood, and environmental and sanitary rules have been enforced for avoiding contamination by human pathogens and hazardous chemicals (see for example EC regulations 853/2004 and 2073/2005). In deltaic lagoons, primary productivity is also a key issue, these ecosystems being exposed to the nitrogen and phosphorus loadings delivered by rivers which drain heavily exploited watersheds.

In the Po river delta and, more generally, along the Northern Adriatic coast of Italy, lagoons tend to be eutrophic and often hypertrophic (Abbiati *et al.* 2010). These conditions are detrimental for preserving ecosystem and water quality, eutrophication being a degenerative process which often leads to hypoxia, or even anoxia, mass fish kills, and biodiversity loss (Cloern *et al.* 2001). On the other hand, shellfish species feed on phytoplankton and suspended particulate matter, therefore and within certain limits, an high primary productivity, which is typical of eutrophic waters, is suited for a better shellfish production (Naylor *et al.* 2000; Newell, 2004). However, over a given threshold, eutrophication leads to organic matter accumulation, which fuels bacterial decomposition processes and oxygen consumption (Viaroli *et al.* 2010a). In turn, oxygen shortage is extremely dangerous for clams themselves which can suffer for mass mortality.

Water quality depends upon several factors which affect the lagoon at different spatial and temporal scales. Clam farming is the main local factor, which impacts directly the lagoon ecosystem: for example natural benthic fauna is suppressed by farmed species and benthic vegetation is eradicated by sediment dredging during clam harvesting (Castaldelli *et al.* 2003; Viaroli *et al.* 2003, 2010a). Further pressures act in the watershed surrounding the lagoon, especially agriculture which delivers fertilizers and pesticides into the lagoon, where they can accumulate in sediment and living biomass (Viaroli *et al.* 2006; Carafa *et al.* 2007). At a wider scale, the lagoon is affected by pressures that originate in the Po river basin. The Sacca di Goro watershed is man-managed for hydraulic safety and agriculture. From September to the next April, canals are kept empty as a prevention measure against flooding, whilst in late spring and summer, water is withdrawn from the main Po river and delivered into a network of canal for irrigation purposes. After irrigation, water is then released into the lagoon. This water is thus contaminated by fertilizers and pesticides from local sources, but the Po river itself accounts for a background pollution which cannot be managed locally and avoided with only local intervention (Viaroli *et al.* 2006).

Climate changes are expected to cause modification of the river and coastal water hydrology (Eisenreich, 2005; Tibaldi *et al.* 2010). Often, periods of very low water discharge are followed by flash floods (Naldi *et al.* 2010), although the long term variability of the Po river water regime is likely multi-factorial (Zanchettin *et al.* 2008). However, in dry years the main effects of pulsed hydrological conditions are the saline wedge ingression at low discharge rates, alteration of nutrient stoichiometry, and modification of aquatic communities and food webs (Naldi *et al.* 2010; Viaroli *et al.* 2008).

In the last two decades, several measures have been undertaken by local authorities and regional environmental agencies in order to mitigate impacts and to avoid direct pressures. Most of the intervention were carried out within the lagoon, often as remedial measures of local degenerative processes. For example, the sand barrier was cut during anoxic crises in order to improve water exchanges with the adjacent sea and to contrast anoxia, as well as channels were dredged within the lagoon to improve water circulation and renewal. At the wider scale of the Po river watershed, environmental policies have been enforced for improving water quality and for achieving the objectives of the Water Framework Directive. Nonetheless, the whole watershed of the Po river is still under threat, due both accidental events, that is the oil spill in February 2010, and diffuse nitrate pollution mainly from agriculture and livestock (Bartoli *et al.* 2012).

The Sacca di Goro lagoon belongs to the Regional Park of the Po River Delta. Nature protection sites are under strict rules for protecting endangered habitats and species, which not necessarily address a better water quality. Moreover, these rules are often perceived too strict and as detrimental for lagoon exploitation, especially for clam farming.

Overall, environmental, economical and social issues are tightly interlinked and their balance is often a critical point in order to avoid conflicts among different groups of stakeholders, mainly at the local scale, and for pursuing either exploitation or conservation goals.

This small but complex ecosystem can be viewed as a metaphor of the physical bottleneck which connects watersheds to the adjacent coastal areas. Concurrently, a bottleneck can be found for policy and decision making, in between the wider scale (region/river basin) and the local entity (municipality/lagoon and its surroundings). In this context, gaps are also in between policies and decisions from one side and their perception by citizens from the other side, and between scientific knowledge and its perceptions by policy and decision makers, end users and citizens.

8.2 GEOGRAPHICAL CONTEXT

8.2.1 The regional scale: Po River and its watershed

The Po River drains a large part of Northern Italy with a surface of 71,000 km^2 (about one fourth of the surface area of Italy) and 17 millions of inhabitants (about one fourth of the total population of Italy). The economical activities in the watershed accounts for approximately 40% of the Italian Gross Domestic Production (Viaroli *et al.* 2010b). Concurrently, a great nutrient and pollutant loading is delivered into the Adriatic Sea, favoring both eutrophication and contamination. The Po River delta has a surface area of ca. 3000 km^2 and composes of seven main deltaic branches and a dozen of coastal lagoons (Figure 8.1).

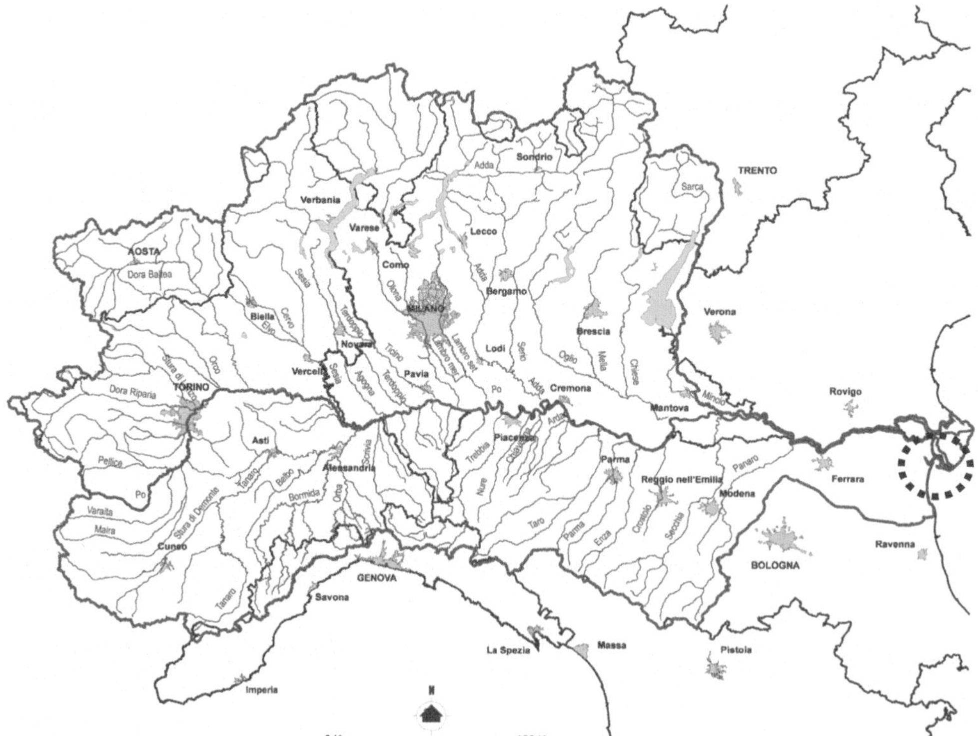

Figure 8.1 Map of the Po river watershed (dark contour) with indicated the main rivers and streams and the Sacca di Goro lagoon within the dashed circle (the map was kindly provided by the Authority of the Po river basin).

8.2.2 The local scale: Sacca di Goro lagoon

8.2.2.1 General features

The Sacca di Goro (44.78–44.83°N and 12.25–12.33°E) is a relatively small (26 km^2) and shallow (mean depth 1.5 m) lagoon located in the Southern part of the delta, in the Province of Ferrara (Figure 8.2).

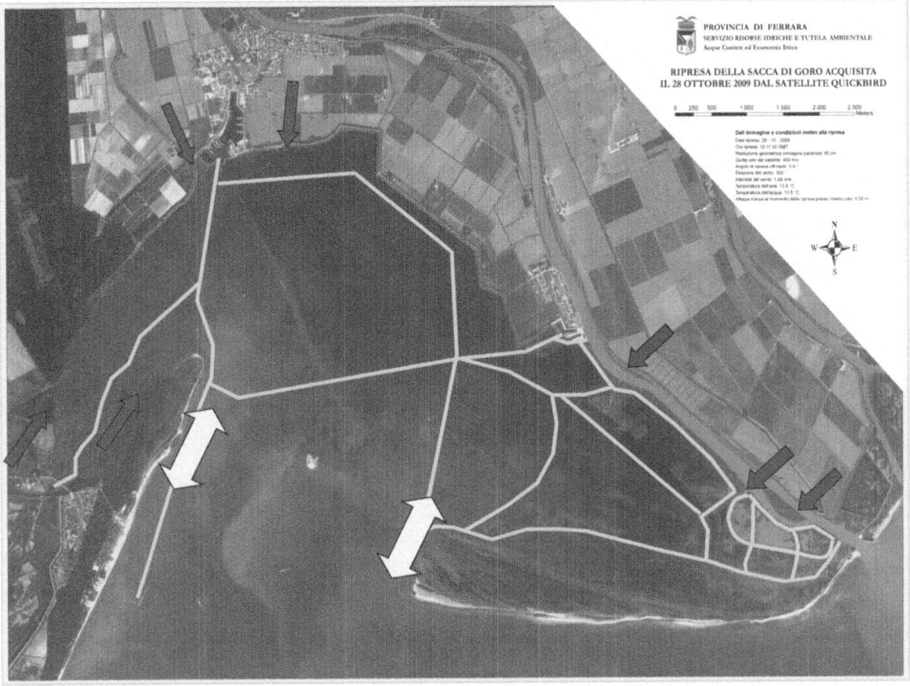

Figure 8.2 Map of the Sacca di Goro lagoon with the main freshwater inputs (dark arrows) and exchanges with the Adriatic Sea (light arrows). The light lines indicate the main navigation channels.

The lagoon watershed extends over 860 km^2 and is partially subsident and man- regulated. The lagoon receive freshwater inputs from the Po di Volano and a series of scooping plants which drain the lagoon surroundings. Additional freshwater inputs are from the Po river through the Po di Goro branch and the river plume which enter into the lagoon from the Adriatic Sea. The lagoon is eutrophic and from mid 1980s abnormal blooms of green macroalgae from the *Ulva* complex have been occurring with frequent dystrophic outbreaks in summer. The main economical activities in the study area are clam farming in the lagoon and agriculture in the terrestrial part of the basin. Industry and urban areas are less developed and population density is relatively low, with a negative trend in the 1991–2001 decade, that is typical of rural areas in Italy. More details and references on the lagoon hydrology, ecology and pollution are reported by Viaroli *et al.* (2006; 2008; 2010a).

8.3 AVAILABLE DATA AND SCIENTIFIC BACKGROUND

The scientific relevance of this small aquatic ecosystem depends upon the number of research programs that have been realized here since 1987 (Table 8.1). The Sacca di Goro lagoon has been an experimental site for analyzing interactions among river basins and adjacent coastal zones. Several international, national and local projects have been carried out dealing with water quality and eutrophication phenomena, ecosystem structure and processes, aquatic resource exploitation, relations between anthropogenic activities, impacts and preventive and remedial measures, scientific support to management and shellfish farming.

Ecological and biogeochemical investigations were conducted in the framework of five EU-funded projects, among which CLEAN, ROBUST and NICE projects provided background studies on the main ecological features, buffering and carrying capacities of the lagoon ecosystem. Ecological and socioeconomic investigations were developed in an integrated manner in the DITTY project. The main outcome of the DITTY project was a Decision Support System, which was developed as a prototype tool for coastal lagoon management in the Southern European Arc.

Additional data were collected during several national and local projects which focused on ecological, geological and socio economical aspects. In most of these projects, the Province of Ferrara was the local coordinator or one of the main actors, through the Operational Unit on Coastal Waters and Fishery (OU). This OU has implemented an extensive monitoring system of oceanographic measures and an early warning service for preventing anoxia effects on farming areas. The OU also provides thematic GIS maps and other informational tools for the planning and management of the lagoon and coordinates water management and dredging activities in the lagoon with the support of the fishermen consortia.

Table 8.1 Main project funded by the European Commission (EC), Italian Ministry programs (IT) and province of Ferrara (FE) which have dealt with the Sacca di Goro lagoon.

Studies of N- and P-cycles and eutrophication in the deltas of the rivers Ebro, Po and Rhone	EV4V-0131-I	EC	1989–1992
Coastal lagoon eutrophication and anaerobic processes (CLEAN)	EV5V-CT92-0080	EC	1993–1994
The role of buffering capacities instabilising coastal lagoon ecosystems (ROBUST)	ENV4-CT96-0218	EC	1996–1999
Nitrogen cycling in coastal ecosystems (NICE)	MAS3-CT96-0048	EC	1996–1999
Development of an Information Technology Tool for the Management of European Southern Lagoons under the influence of river-basin runoff (DITTY)	EVK3-CT-2002-00084	EC	2003–2006
Production, change and vulnerability of transitional aquatic environments along the Adriatic coast	COFIN 1999 (Italian Ministry of University and Scientific and Technological Research)	IT	1999–2001
Ecological integrity indicators of Adriatic transition environments	PRIN 2003 (Italian Ministry of University and Scientific and Technological Research)	IT	2003–2004
Impact of bivalve filter feeders farming on nutrient recycling in dystrophic coastal lagoons: identification of managing strategies for clams farming and the control of algal blooms with low environmental impact	MIPAF 6-C-76 (Italian Ministry of Agriculture and Fishery)	IT	2003–2004
Integrated studies of the Ecology of the Sacca di Goro	Province of Ferrara	FE	1987–1991

Data on water flows and water quality are available since 1990 from the regional Agency for Environmental Protection (ARPA Emilia Romagna) along with data of sanitary concern, provided by the Local Sanitary Agency (ASL Ferrara). Socio-economic data are available from census data of the National Institute of Statistics (ISTAT), and from regional services and local consortia.

8.4 SOCIO-ECONOMIC CONTEXT AND ISSUES

8.4.1 Clam farming

The lagoons of the Northern Adriatic Sea accounts for the production of 50,000-60,000 t yr^{-1} of the Manila clam (*Ruditapes philippinarum*), which represent approximately the 70% of the European clam production. In the Sacca di Goro lagoon clam farming was started in 1986, becoming the main economic activity five years later. Actually, about half of the lagoon surface (13 km^2) is exploited (Figure 8.3), with an annual production oscillating between 15,000 and 18,000 t yr^{-1} (Figure 8.4). Despite its small dimensions, the Sacca di Goro lagoon supports the local economy and provides the main revenue of the resident population, which ranges between 50 and 100 M€ yr^{-1}. Shellfish farming is socially relevant, it ensuring about 1,500 direct job positions, plus employments in seafood industries, commercial and side activities (e.g. shipbuilding). Clam farming is managed by cooperatives of fishermen that access licensed areas, under the control of regional and local authorities.

The extension of licensed areas and the product quantity which is daily delivered to the market are controlled by the cooperative consortia in order to guarantee both food quality and adequate revenue. For this reason, each fisherman can harvest a fixed quota that is established day by day, based on the market demand. Harvesting is performed with manual dredging, which causes sediment alteration and resuspension with local effects on water quality (Viaroli *et al.* 2003). The deterioration of water quality has been also claimed as one of the major causes of macroalgal blooms which are often causing serious setbacks due to massive clam mortality, especially in summer.

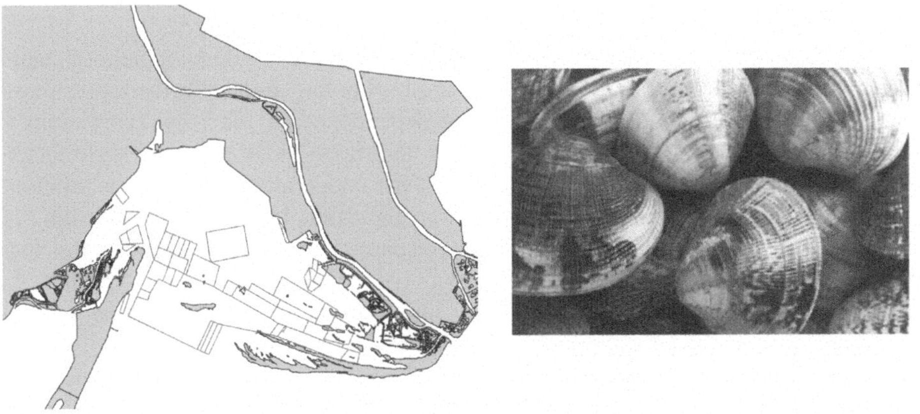

Figure 8.3 Left panel: map of the Sacca di Goro lagoon with licensed areas (grey lines) which are exploited for clam farming. Right panel: Manila clams (*Ruditapes philippinarum*) at commercial size (30–40 mm).

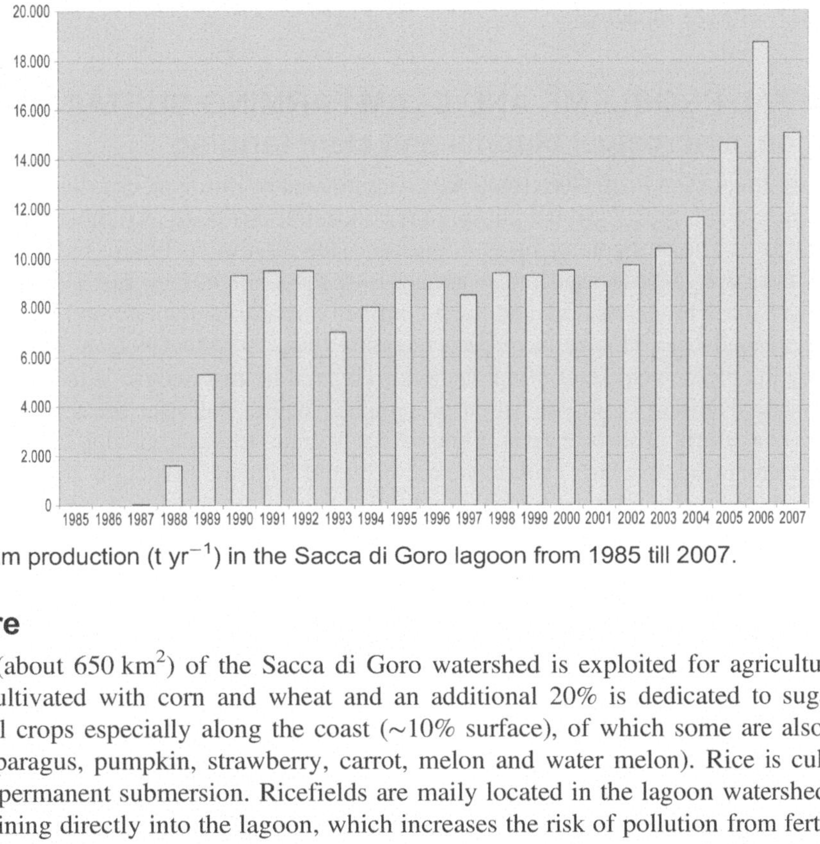

Figure 8.4 Annual clam production (t yr^{-1}) in the Sacca di Goro lagoon from 1985 till 2007.

8.4.2 Agriculture

Approximately 80% (about 650 km^2) of the Sacca di Goro watershed is exploited for agriculture. About 40% of the agricultural area is cultivated with corn and wheat and an additional 20% is dedicated to sugar beets and soybean. Vegetables are typical crops especially along the coast (~10% surface), of which some are also relevant products for export (e.g. salad, asparagus, pumpkin, strawberry, carrot, melon and water melon). Rice is cultivated in paddy soils (~12% surface) with permanent submersion. Ricefields are maily located in the lagoon watershed and are connected to the canal network draining directly into the lagoon, which increases the risk of pollution from fertilizers and pesticides.

8.4.3 Nature conservation and tourism

The Sacca di Goro lagoon belongs to the Regional Park of the Po River delta, under the administration of the Emilia Romagna region. The park is about 600 km^2 and includes part of the ancient and new delta and aquatic and terrestrial environments along the coast (http://www.parcodeltapo.it/). The natural park composes of a variety of habitats and biotopes under the rules of Ramsar convention, UNESCO heritage sites, and Natura 2000 Network. Along the North-Western margin of the lagoon there is a residual Mediterranean forest (Gran Bosco della Mesola).

The lagoon is also surrounded by historical sites: the Pomposa Abey with the old monastic complex (6th century A.D.), the Castle of Mesola, the former hydraulic defenses that were designed in the 16th century A.D., the town of Comacchio and the Etruscan archaeological remains of Spina.

Paradoxically, the presence of valuable landscape and naturalistic areas are not perceived as a development opportunity, but they are rather considered as constraints for the lagoon exploitation. Despite the great naturalistic and historical values, tourism activities are not developed, if not along the coast south of the lagoon, where low cost beach activities are well developed. The naturalistic tourism and eco-fishery are at an early stage as most of the resident workers are mainly employed in clam farming activities.

8.4.4 Social and economical issues

The Sacca di Goro lagoon lies in the Goro municipality. Since mid 1980s, the fast development of clam farming with great incomes has caused a series of social problems, first of all a de-schooling trend. Furthermore, most of the traditional activities, e.g. capture fishery, were abandoned causing the dependence of the local community on the only clam farming. This economic structure is per se fragile and depends both upon local and wide scale factors. The main point concerns the effectiveness of the fishermen organization in cooperatives and collaboration/conflicts among cooperatives. This kind of organization requires a coordination, for example either for controlling harvest and market price of clams, or for managing the lagoon as a common resource. The vulnerability of the local context can be countered by the differentiation of economic activities, for example. with the introduction of other farmed species (e.g. oysters) or with improvement of touristic activities. Overall, the main point deals with how social conditions can be ameliorated, with particular reference to education and reduction of social conflicts.

When shellfish farming is perceived for its positive economical value, less is known about its negative impacts on water and ecosystem quality. Even more, scarce or null awareness is posed to its negative influences on the social system, since it is orienting most of the workforce to shellfish farming, which depress other economic initiatives.

The combination of scarce ecosystem quality with poorly differentiated economic activity and deteriorated social context are symptoms of both socio-economic and environmental vulnerability, as well as they are indicating that the local system is loosing its resilience capacity.

8.5 ENVIRONMENTAL PROBLEMS AND CLAM FARMING SUSTAINABILITY

8.5.1 Eutrophication, macroalgal blooms and clam farming

Freshwaters from the Po di Volano and Po di Goro rivers are contaminated by nitrogen, especially by nitrates, which in late winter and spring attain up to 5 mg N L^{-1}, whilst phosphorus concentrations peak in summer up to 500 μg P L^{-1}, with average values ~100 μg P L^{-1}. The nitrogen loading has been considered responsible for the abnormal growth of green nitrophilous seaweed of the *Ulva* complex and red macroalgae of the *Gracilaria* genus (Viaroli *et al.* 2006, 2010a). Usually, macroalgal growth starts in the shallow and most confined areas in late winter and the maximum spreading is attained in late spring. Extended *Ulva* blooms have been recorded from 1987 to 1992, in 1997–98 and more recently from 2005 to date. During the bloom peak, half of the lagoon is covered by macroalgae with biomass densities attaining 400–800 g m^{-2} as dry matter. Drifted biomasses often accumulate in the shallower areas, especially in the southern zone where clams concessions form a sort of barrier (Figure 8.5). In June–July, the sudden collapse of macroalgal mats is followed by biomass decomposition which causes anoxia, bacterial sulphate reduction and sulphide release into the water column. During neap tides and under calm wind conditions, anoxia persists for several days, causing crop losses up to 30% of the annual production.

Figure 8.5 Macroalgal mats (*Ulva* complex) in the eastern zone of the Sacca di Goro (Upper panels). A large *Ulva* thallus and macroalgal harvesting (lower panels).

In order to either avoid or mitigate the impact of blooms, macroalgal biomass are harvested when attaining thick mats which cause serious threats to clams (Figure 8.5). However, the effectiveness of macroalgal harvesting seems to be low, because the harvested biomass is only a small fraction of the whole biomass bulk. This is an interesting point in terms of better linking science and management. Studies were performed in the lagoon demonstrating that to be effective the harvesting rate has to be much greater than macroalgal growth rate, which can be achieved with an optimization of the cost to benefit ratio (Cellina *et al.* 2003). Nonetheless, macroalgal harvesting is generally designed without taking into account the scientific outcomes.

Shellfish species are filter feeding on phytoplankton and they control its growth. At the farmed animal densities metabolic rates account for high oxygen consumption and release of inorganic nutrients and faecal material. In sheltered lagoons, these processes can generate a feedback loop with phytoplankton and macroalgae (Viaroli *et al.* 2010a). Both field and laboratory experiments demonstrated that clams regenerate ammonia at greater rates than phosphates, thus changing nutrient stoichiometry and potentially favouring nitrophilous macroalgal species (Bartoli *et al.* 2001; Nizzoli *et al.* 2007).

Clams are infaunal species, therefore they rework the sediment and can potentially induce reoxidation processes. However, in the farmed areas, animal densities are very high and respiration processes can contrast sediment reoxidation. Along with metabolic processes, the clam harvesting induces profound changes in sediment biogeochemistry. In fact, clam are collected by either manual or mechanical dredging, which rework the upper 20–30 cm sediment horizon (Viaroli *et al.* 2003). Sediment is thus altered with short term detrimental effects, namely resuspension of the finest silt fraction, increased oxygen uptake and release of reducing ions. In the long term, such sediment wash-out may provide sediment recovery.

8.5.2 Sustainable clam farming

The awareness of the relationships between ecosystem status and exploitation limits, and their dependence on feedback loops among clams, phytoplankton and macroalgae are not yet fully achieved. Nonetheless, the concurrent deterioration of the water quality and crop losses ultimately poses the key question of how to combine exploitation with preservation of environmental quality. These issues have been addressed within the DITTY project with modelling and a decision support system tools (Marinov *et al.* 2007a,b; Mocenni *et al.* 2009).

Clam productivity depends upon hydrodynamics, salinity and freshwater delivery from rivers. Calm waters favor spawning and larval stages development, freshwaters provide particulate organic matter and nutrients for phytoplankton growth which are food for clam, whilst exchanges with the adjacent sea supply oxygen and avoid the anoxia risk. Deltaic lagoons are naturally affected by solid transport from river mouths which cause the formation of littoral arrows and/or barrier island. Accordingly, the Sacca di Goro lagoon tends to be closed by the accretion of littoral arrows which avoid water exchanges with the adjacent sea. Therefore, the lagoon-sea mouths have to be managed by dredging and removing newly deposited sediments. Along with mouth maintenance, canals within the lagoon allow a better water circulation and can improve water exchanges. Overall, although expensive, such management interventions are appreciated by all the actors playing in the lagoon. There is the perception that other factors which affect water quality originated in the Po river basin and cannot be controlled with only local interventions. This is a contrasting issue, because the Po river is at the same time the main food source for clams and the main pollution cause for the lagoon.

Recent studies and monitoring activities demonstrate that the Po river delta and its lagoons are under the global change influence. For example, in the very dry period 2003–2007, the reduction of freshwater inputs due to much lesser wet precipitation and more water usage in the watershed caused saline wedge ingression with salinity increase. Moreover, the lesser freshwater discharge also caused a decrease in solid and nutrient inputs which lowered both primary productivity and shellfish production. The lagoon watershed is subject to subsidence, with areas located some meters below the sea level. The combination of subsidence and saline wedge ingression in dry years is compromising natural communities and water uses, for example, for irrigation and for human settlements. Mitigation measures need to be considered for an integrated and sustainable management.

To sum up, the persistence of clam farming is tightly linked to sustainable farming practices and good ecosystems status, which require at least four major components:

- Economy, which deals with clam productivity, quality and marketing;
- Integrated policy of watershed and lagoon management, delivery and control of licensed areas for clam farming, preservation of natural unexploited areas;
- Structure and organization of the local society and citizen participation in the decision making;
- Control and improvement of the environmental status, including water, sediment and biological elements.

In the DITTY project, the identification and prioritization of scenarios was implemented with an approach that integrate ecological and economical factors (Viaroli *et al.* 2007). Basically, this approach allows the identification of the major drivers and their changes in time and space. The scenario analysis is based upon the consideration of the main economical activities in the watershed-lagoon system and information have been organized within a DPSIR scheme for both watershed (Table 8.2) and lagoon (Table 8.3). Here economic and environmental variables have been accommodated in a comprehensive representation of the watershed and the lagoon that describes major interactions between selected compartments.

Table 8.2 Application of the DPSIR scheme to the Po river watershed.

Driver	Pressure	State	Impact	Response
Population Urbanization Infrastructures	Water extraction Wastewater release Soil compaction and impermeabilization	Water quantity and quality Sediment quantity and quality	Water and sediment pollution Community shift, Reduced infiltration	Management of water extraction and uses River restoration Implemetation of wastewater treatment systems
Agriculture Livestock	Fertilization Manure delivery Irrigation Runoff Wastewater delivery	Water quantity and quality Sediment quantity and quality	Water and sediment pollution, Eutrophication, Community shift	Changes in agricultural and animal husbandry practices
Climate changes	Changes in rainfall frequency and quantity	Water level River Discharge Current velocity Increased runoff Erosion	Saline wedge Flash floods Alterations of nutrient stoichiometry	Levees and banks Riparian areas restoration Cover crops

Table 8.3 Application of the DPSIR scheme to the Sacca di Goro lagoon.

Driver	Pressure	State	Impact	Response
Watershed exploitation	Freshwater inputs N and P loadings	Water and sediment quality High biodiversity	Macroalgal blooms Hypoxia, Anoxia	Reduce nutrient loads Macroalgae harvesting Canal dredging
Aquaculture	Sediment dredging Eradication of benthic vegetation Mono-specific benthic communities (clams)	Water and sediment quality High biodiversity	Community shift Pollution Sediment losses Eutrophication Hypoxia, anoxia	Changes in farming techniques Diversified economic activities Canal dredging
Climate changes	Changes in freshwater discharge increase in sea level	Water level River Discharge Runoff	Community Changes Pollution Sediment transport Flooding	Levees and banks Riparian areas restoration Canal dredging

In the DITTY Projects, the Sacca di Goro lagoon has been analyzed with models, indicators, and multi-criteria analyses. A prototypical decision support system (DSS) has been also developed for integrating all the models and indicators in order to better support the decision makers (Mocenni *et al.* 2009). The proposed DSS structure allowed for the integrated system approach, and DPSIR cause-effect relationships has been directly represented in it. It is important to note that all the models that have been developed can be used separately or integrated in the DSS. In particular, the multi-criteria analysis tool can be easily applied to a broad set of decision problems and it is very intuitive to understand and practice. In any multi-criteria analysis, the final step of ranking and selection of the alternatives is not a scientific, but a political exercise of negotiation and trade-off of necessarily subjective values and believes, which again makes the direct involvement of end-users mandatory to reach meaningful and practical results. In the DSS, the Analytic Hierarchy Process (AHP) is applied as an effective tool to make decisions in situations where multiple and conflicting criteria are present, and both qualitative and quantitative aspects of a decision need to be considered. AHP is an effective tool for dealing with complex decision making, and may aid the decision maker to set priorities and make the best decision. The computations made by the AHP are always guided by the experience of decision makers experience, and the AHP can thus be considered as a tool that is able to translate the evaluations into a multi-criteria ranking.

The main results of the exercise from the DITTY project indicate that both increase or more intense exploitation of the already licensed areas in the lagoon are not sustainable and either the risk of anoxia or the ecosystem vulnerability to Ulva blooms will increase (Viaroli *et al.* 2007). Only the sea-side area of the lagoon could be exploited, without increasing the risks of anoxia and the vulnerability to blooms. Furthermore, the cost of the management measures that have to be undertaken for controlling macroalgal blooms are potentially compensated by the increase in clams production.

The focus on clam farming provides only a partial view of a more complex system which composes by agriculture, urban, touristic areas, and natural sites and is under the influence of climate changes. However, the scenarios depicted in the DITTY studies constitute only the first stage in the procedure of overall lagoon scenario analysis, and a second iteration is necessary to define the main constraints and boundaries, as well as to incorporate the economic evaluation and social implications of the different options.

8.6 THE AWARE EXPERIMENT ON HOW CONNECTING PEOPLE TO SCIENCE AND DECISION MAKING IN THE SACCA DI GORO

8.6.1 State-of-the art information provided to participants

The information provided in the previous sections identified the Sacca di Goro lagoon as a fragile and highly impacted system. The more effective way to maintain and improve water quality in the lagoon in the short time is to maintain and improve water exchanges with the adjacent sea. For this reason, the main management options undertaken to date have mainly addressed hydrodynamics with local intervention, for example canal dredging. However, most of the nutrient loads come from the Po river basin and are not directly related to the local activities. The application of the EU directives to the whole basin is expected to drive a consistent nutrient reduction but recent estimates indicate that the deltaic lagoons will likely maintain an high eutrophication levels (Provini & Binelli, 2006).

Other works are planned to dredge channels and maintain the sea mouths open. The last aspect is relevant since the solid transport from the Po river is continuously adding material on the sand banks, thus reducing the sea-lagoon water exchanges.

The macroalgal blooms are a serious environmental problem since the huge accumulation of easily degradable biomass can strongly affect biological communities, including clams. After the bloom peak and during the following decay phase, the water would become anoxic imposing highly reducing conditions, which are a serious threat for benthic and fish fauna. The macroalgal growth is linked not only to the nutrient availability but also to meteorological conditions, storminess and hydrodynamics, the blooms being favoured by water stagnation and high temperature. Finally, feedbacks loop among clams, phytoplankton and macroalgae have been invoked as a possible cause of shifts from phytoplankton-dominated to macroalgal-dominated conditions. Namely, clams can remove phytoplankton, increment mineralization rates of dissolved organic nutrients, improve water transparency and provide hard substrata for the first stages of macroalgal growth.

Actually, the more efficient way to prevent macroalgal blooms is the harvesting in an early growth stage. Fishermen with the support of local authorities organize boats which can effectively remove macroalgal biomasses only when density is low. When the biomass density is high, the harvesting is completely ineffective as growth is density limited and the removal of macroalgae would only stimulate further blooms.

Sacca di Goro is considered sensitive to global climate changes due to its position at the sea-land interface. The scenarios identified for the Mediterranean area highlight a possible reduction of water availability, increment of drought, severe biodiversity loss, reduction of suitable cropping areas, increased land loss, increased salinity and eutrophication in estuaries and deltas (http://www.ipcc.ch/publications_and_data/publications_and_data_reports.shtml).

These scenarios challenge the management of the Sacca di Goro in which large parts of the watershed is already below the sea level and the main economic activities are affected by salinity.

8.6.2 The experiment: a cooperative decision system

Within the AWARE project, the citizen panel was involved along with scientists, stakeholders, end-users and policy and decision makers working in the lagoon (Giordani *et al.* 2010). The first step was the scientific dissemination to citizens, with presentation and discussion of the main scientific background on this system.

The citizen panel then reassessed the scientific knowledge, in the light of their background, own needs and perceptions of policies and exploitation of the lagoon. Finally, a document was delivered that included their own recommendations.

From the scientific communication they perceived mainly issues related to the local dimension, which can be directly experienced, namely water quality, clam farming and safe environment. However, they were also aware of factors playing at a wider scale, especially human activities in the Po river watershed, climate changes and sea level rise, which are currently reported by the media, most often with partial views and biased information. Ultimately, the actions citizen have proposed come basically from the day-to-day experience rather than from the scientific knowledge.

The citizens posed to scientists working in the lagoon a series of questions dealing with the perception of environmental dangers and how they can contribute to problem solving. Four main categories of problems were identified, namely the impacts of Po River on the lagoon, the key role of clam farming in the local economy and societal structure, possible alternatives to the current business, and the global change scenario (Table 8.4).

Table 8.4 Main points identified by citizens following scientific reporting and communications.

Macroalgal blooms depend on the nutrient loading delivered by the Po river Management of macroalgal blooms requires attention for biomass fate and possible uses The solid transport from the Po river, can cause siltation, which has to be prevented
Clams are not indigenous species, but were introduced. If appropriate they can be substituted with other farmed species (e.g. oysters) or activities (e.g. tourism) Clam farming is regulated by strict rules. Preventive measures for maintaining and implementation strategies for ameliorating such activity have to be adopted Clam farming is a monoculture, one can expect a decrease of genetic diversity with possible weakening of the species
The increasing income from clam farming was not accompanied by the improvement of education and welfare, which can be a cause of future societal problems Activities in the lagoon and in its surroundings can generate conflicts among different stakeholders Clam farming is not compatible with conservation priorities, namely it cannot be performed in the Natura 2000 zones Alternative activities have to be identified, for example tourism
The lagoon surroundings are mainly below the mean sea level, which could potentially amplify the foressen effects of sea level rise The main point is to identify the fate of the lagoon ecosystem under so many threats.

To sum up the key topics identified and deepened in the local workshop addressed few major points.

(1) The vulnerability of the Lagoon is caused by Po River pollution. Therefore, in order to improve water quality in the lagoon measures have to be undertaken upstream and not only in the lagoon.
(2) The pollution loading from agriculture in the lagoon surroundings has to be reduced with preventive measures and better agricultural practices.
(3) Alternative economical activities have to be developed, through the exploitation of protected areas and natural habitats. The great natural value of these zones can give rise to a sustainable eco-tourism, which in turn can support local economy by purchasing local goods and participating with small local businesses.
(4) Information of natural goods and possible threats of this territory are often fragmented, which results in a poor governance.
(5) There is a need of applied research and technological innovation as a tool for supporting management and economy of the area. One example can be given by the use of the macroalgal biomass, for example, for fuel production. The support of the scientific community would be important here.

In contrast with the above general statements, the proposed actions were still identified locally, basically having environment protection and economy development as core issues (Table 8.5). In the first case, citizens recommended to undertake rapid actions to improve water quality in the lagoon, to promote collaboration among authorities and research institutes. They suggested toenforce regional and national rules, in order to have long term preventive measures at the river watershed scale for improving the sustainability of the human activities in the whole watershed, with cascade benefits also at local level. The citizen panel underlined also the needs to have effective and objective research activities coupled with transfer of results in an understandable way to decision makers and citizen, in order to improve connectivity among these actors and develop appropriate management options, in which citizens needs are incorporated.

Table 8.5 List of priority options identified by citizens.

PO1	Structural and hydraulic interventions in the Sacca di Goro lagoon for improving water quality
PO2	Installation of new wastewater treatment plants in the Po river basin, including diffuse phyto-treatment systems
PO3	Ameliorating the surveillance in the lagoon (e.g. monitoring of clam concession areas) and in the Po river (e.g. contamination monitoring)
PO4	Raising fishermen awareness of environmental impacts of their activities
PO5	Strengthening cooperation among actors and institutions involved in the lagoon management
PO6	Promoting environmental tourism (e.g. with communication campaigns)
PO7	Disseminating information and awareness on the lagoon status
PO8	Sensitizing local communities and authorities on the importance of solid waste management

The outcomes of the local participatory process have been integrated with the application of two multicriteria analyses for decision making. The Analytic Hierarchy Process (AHP) has been applied as an effective tool to make decisions in situations where multiple and conflicting criteria are present, and both qualitative and quantitative aspects of a decision need to be considered. A more general analysis has been carried out by means of the Analytic Network Process (ANP) method which considers the system as a network. The ANP methodology takes into account the influence of Benefits, Opportunities, Costs and Risks in the problem (the methodology is also named BOCR analysis).

Both AHP and ANP are based on pair-wise comparisons of options and criteria in order to measure the weights of each component, and finally provide a prioritization (AHP) and influence weights (ANP) of the alternatives available in the decision.

The alternatives listed in Table 8.5 have been re-assessed through interviews of stakeholders, policy makers, and scientists considering three criteria: sustainability and environmental quality (environmental criterion), Cost-Benefit evaluation (economic criterion) and life quality (social criterion). From AHP application PO5 and PO7 result the first priority; PO1, 2, 3, 4 and 6 are grouped in a second cluster and PO8 is the last option. Therefore, collaboration among actors and information were considered as the most urgent priorities.

For applying the ANP method priorities in Table 8.5 have been grouped into three main cluster, based on option similarities: PO5 and PO7 are identified as information and communication priorities (B1), PO1 and PO2 are considered as structural and hydraulic options (B2), and PO3, PO4, PO6 and PO8 are assumed as development and control options (B3). ANP outcomes evidence that B1 is the first priority for all the considered criteria, whilst B2 and B3 differ only for O and C (Table 8.6).

Table 8.6 Priority ranking of alternative according to the BOCR scheme.

Alternatives	Options	Benefits	Opportunities	Costs	Risks
B1	PO5 + PO7	1	0.68	0.75	0.47
B2	PO1 + PO2	0.66	0.88	0.54	0.84
B3	PO3 + PO4 + PO6 + PO8	0.6	0.56	0.89	0.88

Afterwards, the ANP criteria have been combined in three different rules with three algorithms:

$$\text{Rule 1} = B + O + (1/C) + (1/R);\ \text{Rule 2} = B + O\text{-}C\text{-}R \text{ and Rule 3} = B^{*}O/C^{*}R$$

The ranking of alternatives with the application of the different rules evidence that B1, "Information and communication" is the priority (Table 8.7). Moreover, there is a remarkable distance between the first and the second position indicating that results are particularly robust. The alternative B3, "Development and control", reports always the lowest score, whilst the alternative B2, "Structural and hydraulic interventions" is always in the second position.

Table 8.7 Global ranking of alternatives according to the rules reported in the text.

Alternatives	Rule 1	Rule 2	Rule 3
B1	5.12	0.46	1.92
B2	4.58	0.17	1.29
B3	3.42	−0.6	0.43

Overall, the outcomes of both AHP and ANP are in agreement with the final citizen declaration that pointed out how the communication gap among actors is critical and a better knowledge of problems is fundamental for improving both citizen awareness or their participation in problem solving.

8.7 CONCLUSIVE REMARKS – HOW TO FURTHER DEVELOP CITIZEN AWARENESS ON SUSTAINABLE EXPLOITATION OF THE SACCA DI GORO LAGOON AND ITS WATERSHED

The Sacca di Goro case study was discussed within a Policy and Stakeholder Advisory Group (PSAG) focusing on the institutional framework and recommendations to enhance the participatory process. Two important aspects have been highlighted, which represent useful inputs for framing the discussion with the citizens and stakeholders and for further develop a full scale local participatory process.

First and foremost, the management of the Sacca di Goro lagoon would require a better coordination amongst the local decision makers – the Municipality and local consortia – that do take decisions on local matters (e.g. the recent investment for reducing the natural sand barrier and open the lagoon to a greater inflow of sea waters to influence salinity – a key parameter for clam farming) and the Po Basin authority and Emilia Romagna region, that takes decisions which affect the quantity and quality of freshwater in the Po river delta.

Based on the Water Framework Directive (2000/60/CE) the Sacca di Goro lagoon is as a matter of fact a "highly modified water body", due to its position at the outlet of the Po river and the intense clam farming. About 50% of the lagoon surface is indeed man-managed and deeply altered for clam sowing and harvesting. Moreover, the great part of the lagoon, also the most natural areas, are impacted by canal dredging and river regulation. This fact is not sufficiently acknowledged at the policy level and is almost unknown to citizens, whilst would be greatly appreciated by sector stakeholders. The decision on how to classify the lagoon is difficult for policy makers at either regional or national level, because it would require a special legislation. At present, the lagoon is still classified as a natural water body.

The discussion with the citizens could benefit from focusing on how to classify the lagoon for its heuristic meaning. In fact, accepting the lagoon to be a natural ecosystem, implies that it has to attain a "good ecological status" within a given deadline, which at present is fixed at 2027. For achieving this goal, citizens have to be aware that scenarios and options of future development have to be deeply changed, and only cosmetic interventions will be not sufficient. Thus, what identified by them with this project has to be considered as a very preliminary step and a further participatory process has to be launched for undertaking more rules and restriction in lagoon exploitation. Likely, this is a conflicting issue.

The alternative, is to maintain the current intensive clam farming, but accepting that the lagoon is an "highly modified" water body for which only the "good potential ecological status" has to be identified. In this case, the main goal is only to identify which decisions and actions have to be enforced in case one would decide to practically achieve the good ecological status. This latter, could become an interesting exercise, because decision will not have immediate effects, but will be only planned, so conflicts will be likely smoothed. Within this context, different actors and players would have the opportunity to further develop what was early started with the DITTY project and further with the AWARE project. Meanwhile, ecological and socio-economic risks which are associated to the different options could be better evaluated and an adaptive approach could be adopted.

A key point was not completely understood until now is that coastal lagoons are intrinsically unstable. They are also a sort of bottleneck in between terrestrial and marine domains. Among others, deltaic lagoons are especially threatened because they are the recipients of what is delivered by river basins. When the upstream system is heavily impacted, lagoons will have a great vulnerability. In turn, this will result in high fragility and low resilience of the socio-economical system which depends on it.

Furthermore, one can adopt an ecological paradigm which has been greatly underlined in the Millenium Ecosystem Assessment (www.maweb.org), that is the system stability and/or resilience depends upon (bio)diversity. This point has been also perceived by citizens, who proposed to adopt more diversified actions/options.

Finally, we have to take into account that citizens expressed the fear of wasting their time: this can be both interpreted as an expectation of an effective use of time and as they were suspicious of being involved in a process that may produce no results. This feeling poses a key point which is the effective willingness of all actors to participate in the decisional process, avoiding preventive lobbyings.

In all the three participatory moments knowledge has been provided by all the participants, in different formats and measure: expert knowledge mainly by scientists, tacit and local knowledge mainly by stakeholders and local policy makers. The reaction of citizens displayed gradually within the process: their declaration, at the moment, can be thought as the closest objective measure of the knowledge they accumulated so far. The last, but absolutely not least, point is that information and communication are the most relevant constraint in developing a participatory approach to decision in this context. It is a "cultural bottleneck" which tackles mainly scientists and policy makers.

REFERENCES

Abbiati M., Mistri M., Bartoli M., Ceccherelli V. U., Colangelo M. A., Ferrari C. R., Giordani G., Munari C., Nizzoli D., Ponti M., Rossi R. and Viaroli P. (2010). Tradeoff between conservation and exploitation of the transitional water ecosystems of the Northern Adriatic Sea. *Chemistry and Ecology*, **26**, 105–119.

Bartoli M., Nizzoli D., Viaroli P., Turolla E., Castaldelli G., Fano A. E. and Rossi R. (2001). Impact of Tapes philippinarum farming on nutrient dynamics and benthic respiration in the Sacca di Goro. *Hydrobiologia*, **455**, 203–201.

Bartoli M., Racchetti E., Delconte C. A., Sacchi E., Soana E., Laini A., Longhi D. and Viaroli P. (2012). Nitrogen balance and fate in a heavily impacted watershed (Oglio River, Northern Italy): in quest of the missing sources and sinks. *Biogeosciences*, **9**, 361–373.

Carafa R., Wollgast J., Canuti E., Lighart J., Dueri S., Hanke G., Eisenreich S. J., Viaroli P. and Zaldivar J. M. (2007). Seasonal variations of selected herbicides and related metabolites in water, sediment, seaweed and clams in the Sacca di goro coastal lagoon (Northern Adriatic). *Chemosphere*, **69**, 1625–1637.

Castaldelli G., Mantovani S., Welsh D. T., Rossi R., Mistri M. and Fano E. A. (2003). Impact of commercial clam harvesting on water column and sediment physicochemical characteristics and macrobenthic community structure in a lagoon (Sacca di Goro) of the Po river delta. *Chemistry and Ecology*, **19**, 161–171.

Cellina F., De Leo G. A., Bartoli M. and Viaroli P. (2003). Biological and ecological modelling as a tool for supporting macroalgal bloom management. *Oceanologica Acta*, **26**, 139–147.

Cloern J. E. (2001). Our evolving conceptual model of the coastal eutrophication problem. *Marine Ecology Progress Series*, **210**, 223–253.

Eisenreich S. (2005). Climate Change and the European Water Dimension, European Communities, report n. EUR 21553 EN.

Giordani G., Viaroli P., Luè A., Lovo S., Mocenni C., Bencivelli S., Liquete C. and Sessa C. (2010). Sacca di Goro case study content and process design report. AWARE Project, Deliverable D2.3, p. 52.

Marinov D., Galbiati L., Giordani G., Viaroli P., Norro A., Bencivelli S. and Zaldivar J.-M. (2007a). An integrated modelling approach for the management of clam farming in coastal lagoons. *Aquaculture*, **269**, 306–320.

Marinov D., Zaldivar J. M., Giordani G., Norro A. and Viaroli P. (2007b). Integrated modelling in coastal lagoons: Sacca di Goro case study. *Hyrobiologia*, **611**, 147–165.

Mocenni C., Casini M., Paoletti S., Giordani G., Viaroli P. and Zaldívar J. M. (2009). A decision support system for the management of the Sacca di Goro (Italy). In: Decision Support Systems for Risk-Based Management of Contaminated Sites, A. Marcomini, G. W. Suter, II and A. Critto (eds.), Springer Science, New York, pp. 399–422.

Naldi M., Pierobon E., Tornatore F. and Viaroli P. (2010). Il ruolo degli eventi di piena nella formazione e distribuzione temporale dei carichi di azoto e fosforo nel fiume Po (Effects of flooding in the formation and evolution of N and P loadings in the Po river). *Biologia Ambientale*, **24**, 59–69.

Naylor R. L., Goldburg R. J., Primavera J. H., Kautsky N., Beveridge M. C., Clay J., Folke C., Lubchenco J., Mooney H. and Troell M. (2000). Effect of aquaculture on world fish supplies. *Nature*, **405**, 1017–1024.

Newell R. I. E. (2004). Ecosystem influences of natural and cultivated populations of suspension-feeding bivalve molluscs: a review. *Journal of Shellfish Research*, **23**, 51–61.

Nizzoli D., Bartoli M. and Viaroli P. (2007). Oxygen and ammonium dynamics durino a farming cycle of the bivalve Tapes philippinarum. *Hydrobiologia*, **587**, 25–36.

Provini A. and Binelli A. (2006). Environmental quality of the Po River Delta. In: The Handbook of Environmental Chemistry Volume 5, Part H Estuaries, P. J. Wangersky (ed.), Springer-Verlag, Berlin Heidelberg, pp. 175–195.

Tibaldi S., Cacciamani C. and Pecora S. (2010). Il Po nel clima che cambia (The Po river in the climatic change context). *Biologia Ambientale*, **24**, 21–28.

Viaroli P., Azzoni R., Bartoli M., Giordani G., Naldi M. and Nizzoli D. (2010a). Primary productivity, biogeochemical buffers and factors controlling trophic status and eco system processes in Mediterranean coastal lagoons: a synthesis. *Advances in Oceanography and Limnology*, **1**, 271–293.

Viaroli P., Bartoli M., Giordani G., Azzoni R. and Nizzoli D. (2003). Short term changes of benthic fluxes and oxygen depletion risk in a coastal lagoon with clam farming (Sacca di Goro, Po River Delta). *Chemistry and Ecology*, **19**, 189–206.

Viaroli P., Bartoli M., Giordani G., Naldi M., Orfanidis S. and Zaldivar J. M. (2008). Community shifts, alternative stable states, biogeochemical controls and feedbacks in eutrophic coastal lagoons: a brief overview. *Aquatic Conservation: Freshwater and Marine Ecosystems*, **18**, S105–S117.

Viaroli P., Giordani G., Bartoli M., Naldi M., Azzoni R., Nizzoli D., Ferrari I., Zaldívar Comenges J. M., Bencivelli S., Castaldelli G. and Fano E. A. (2006). The Sacca di Goro lagoon and an arm of the Po River. In: The Handbook of Environmental Chemistry Volume 5, Part H Estuaries, P. J. Wangersky (ed.), Springer-Verlag, Berlin Heidelberg, pp. 197–232.

Viaroli P., Marinov D., Bodini A., Giordani G., Galbiati L., Somma F., Bencivelli S., Norro A. and Zaldivar J. M. (2007). Analysis of clam farming scenarios in the Sacca di Goro lagoon. *Transitional Waters Monographs*, **1**, 71–92.

Viaroli P., Puma F. and Ferrari I. (2010b). Aggiornamento delle conoscenze ecologiche sul bacino idrografico padano: una sintesi (Update of the ecological knowledge on the Po river basin: a synthesis). *Biologia Ambientale*, **24**, 7–19.

Zanchettin D., Traverso P. and Tomasino M. (2008). Po river discharge: a preliminary analysis of a 200 year time series. *Climatic Change*, **89**, 411–433.

Chapter 9

Monitoring and evaluation of the AWARE pilot experience

Irina Comardicea[1], Federico Lia[2], Camino Liquete[3], Alessandro Luè[2,4] and Elsa Semmling[1]

[1]*adelphi, Berlin, Germany*
[2]*Poliedra–Politecnico di Milano, Milan, Italy*
[3]*Joint Research Centre–European Commission, Institute for Environment and Sustainability, Ispra, Italy*
[4]*INDACO Department – Politecnico di Milano, Milan, Italy*

9.1 INTRODUCTION

The evaluation of the AWARE process targeted three pilot coastal areas in Europe: the Baltic Sea (Gulf of Riga), the Sacca di Goro (Po Delta) and the Southern North Sea. In all three cases scientific assessments to characterize the ecological status of the three regions and to define policy options for improving management have been undertaken. Some of these policy measures are already under implementation but in all cases it was felt by the scientific and social scientist project partners that more investments and coordination are needed to foster real and sustainable change.

The objective of the AWARE monitoring evaluation was to help project partners, most of whom are scientists, to learn from each other and from their interactions with citizens, stakeholders and policy-makers, with the common goal of developing a more integrative science-policy-public interface. Given that the evaluation team comprised mainly social scientists, this also represented an active incorporation of social science expertise in adaptive ecosystem management, as recommended by the policy and scientific literature.

Past experiences in policy-making participation (e.g. Norton, 2005) have shown that an integrated adaptive approach to ecosystem management is better both in terms of gaining input in a participatory manner, and in delivering policy-relevant output. The AWARE project used this approach to identify and evaluate ways in which citizens' participation could improve the sustainability of coastal water management. This chapter describes and discusses the evaluation undertaken in the project, to assess the success of the tailored knowledge brokerage process used in the three AWARE pilot case studies. Starting with a brief overview of adaptive management research, this chapter describes the evaluation methodology used in the project. This forms the background for a discussion on the results of the evaluation, which is followed by a set of lessons learned from the AWARE evaluation.

9.1.1 Adaptive management experience

As mentioned in the Introduction to this book, AWARE was in a nutshell a pilot experience engaging scientists, citizens, stakeholders and policy-makers in a collective appraisal of coastal water degradation issues, inspired by the concept of adaptive management. This concept was developed first by Holling (1978) and elaborated by Walters and Holling (1990) to describe the importance of both learning and flexible management when addressing ecological problems. It was subsequently used to advocate the use of social science knowledge in environmental research and policy (e.g. Endter-Wada *et al.* 1998; Norton 1998), the inclusion of local or non-expert knowledge (e.g. Castillo *et al.* 2005) as well as the relevance of procedural rules (Gregory *et al.* 2006) and rules for knowledge interfacing and sharing (Roux *et al.* 2006).

The application of adaptive management faces countless challenges (Huitema *et al.* 2009) as well as some scepticism from an operational point of view (Biswas, 2004). However, there are well documented examples of positive lessons learned in adaptive management of natural resources (e.g. Leach & Pelkey 2001; Mostert *et al.* 2007; Allan & Stankey, 2009). According to these authors policies can be changed in response to the knowledge and learning achieved through adaptive processes. Thus, public participation can contribute to decision-making with more creative and new solutions, can improve relations between the actors involved, and can prevent legal disputes during the implementation stage (Huitema *et al.* 2009).

Generally, this paradigm has been further explored (Folke *et al.* 2005) by analysing the capability of a socio-ecological system to reorganise its governance model by connecting individuals, organisations, agencies, and institutions at multiple organisational levels. The complexity of governance in ecosystem management structures cannot be overstated. Figure 9.1 below shows an example of this complexity by focusing on the cause-effect relationships in response to possible interventions. However, the goal of AWARE has been to discuss the EU water policy framework (see Chapter 3 of this book) and how to connect scientific knowledge and water management practices to achieve a good ecological status of the coastal waters. By looking at the extent to which coastal water management is operating in an integrated adaptive management style, AWARE assumes that the science-policy interface in the water sector, as in others, is still unsatisfactory, and that there is a potential for better connectivity between the science and policy groups of actors taking part in this sector.

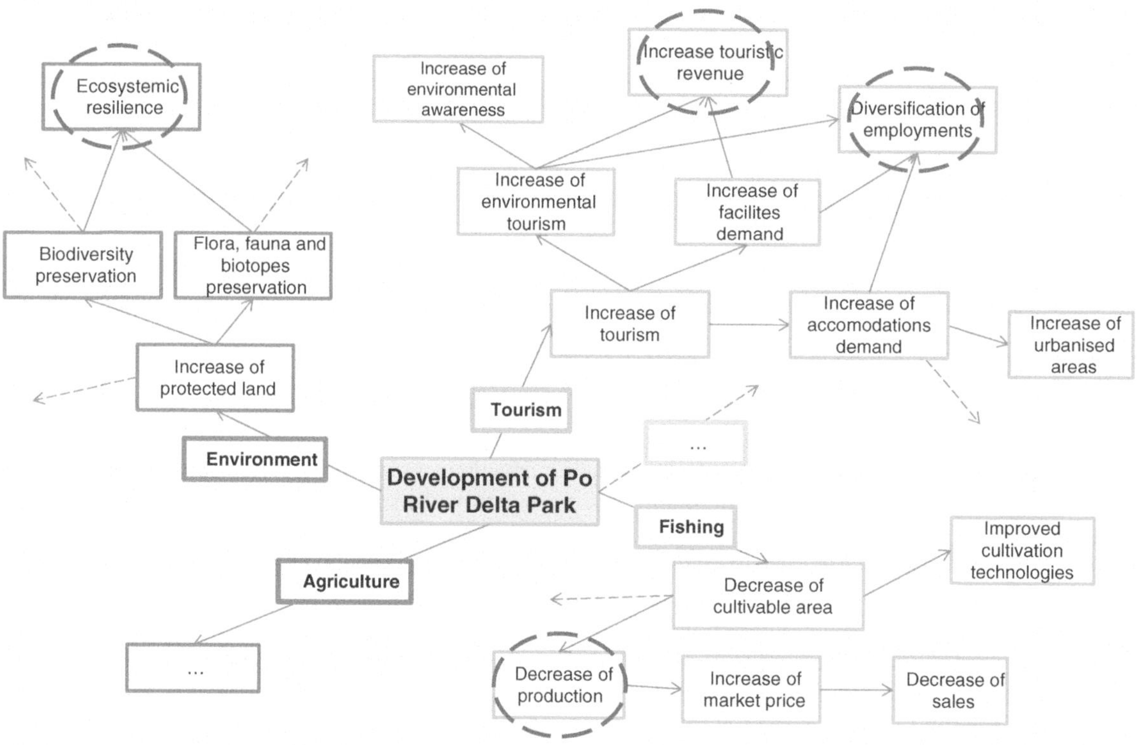

Figure 9.1 An example of systems complexity in terms of cause-effect responses to a possible intervention option, taken from one of the three AWARE case studies – the Po River Delta.

In the case of the water sector, adaptive management practices should increase the adaptation capacity of river basins to changing conditions (Pahl-Wostl, 2007), like the ones resulting from natural or human pressures. A broad body of evidence has been produced in the framework of relatively recent research projects focused on integrated and adaptive water management such as SLIM (Social Learning for the Integrated Managing and sustainable use of water at catchment scale), NeWater (New methods for adaptive water management under uncertainty), HarmoniCOP (Harmonizing COllaborative Planning), or GWSP (Global Water System Project) amongst others. Each of these projects represents a step forward in the integration of society in water and ecosystems management.

Adaptive water management in the European institutions is now being translated into public participation since the EU Water Framework Directive (WFD) introduced in its Article 14 the notion of *public participation* for policy implementation. This Directive considers the participation of stakeholders and the broader public a key element to reduce uncertainty and to improve management (Newig *et al.* 2005). Public participation covers very different meanings such as information supply, consultation, discussions with the public, or co-decision-making (Huitema *et al.* 2009). The WFD considers these as well, along with the active involvement of the public, including users. However, while the WFD and its Common Implementation Strategy (CIS) provide extremely detailed information on monitoring and assessment requirements, the implementation of public participation remains ambiguous and relies on the understanding of each river basin authority. Some examples and analysis of these implementation strategies can be found, for instance, in Borowski and Pahl-Wostl (2008). Member States and river basin authorities usually have limited experience and resources to apply public participation in policy-making. The options of information supply and consultation have been the most widely applied public participation methods, with more or less extent and success. The active involvement however, especially of lay citizens has been very rare, although a few examples are available for instance from the Catalan region of Spain (NN ND).

9.2 EVALUATION METHODOLOGY

Scientists view the policy end-user in research projects as the *legitimate* client for their research results, but on the ground there is a significant lack of transfer mechanisms that would allow passing the relevant information on to the policy-makers and hence, to the public. Therefore, the AWARE project tested a new way of connecting scientists and policy-makers, by inviting a *third player* in the game – citizens interested in, and affected by, water policy decisions – to participate in a knowledge brokerage process.

9.2.1 The AWARE knowledge brokerage process

A detailed description of the tailored knowledge brokerage process used throughout the AWARE project is provided in Chapter 5 of this book, and the monitoring and evaluation discussed in this chapter occurred throughout each step of the project. In short, the project's initial phase started by randomly selecting a panel of ten individuals from three case study areas covering a total of five countries: Estonia and Latvia for the Gulf of Riga case; Belgium and France for the Southern North Sea case; and Italy for the Sacca di Goro case. These citizen panels thus formed a transnational group of 30 members who first met altogether at a 2,5 day European level workshop to discuss their mandate and gain a common understanding of the EU water policy framework and on how to connect scientific knowledge and water management practices to achieve a good ecological status of the coastal waters – the goal of AWARE. Project partners and external speakers were also part of this workshop, thus allowing learning to occur across all actors involved in the project. The project evaluation also began with this workshop by observing both the participatory process as well as the outcomes of the event.

This initial workshop was followed by a set of workshops in each of three case studies, where each citizen panel discussed over two days with a variety of stakeholders and decision-makers the issue of sustainable coastal water management and a good science-policy interface. The evaluation continued at the local events, observing this time also the tangible outcomes of the participatory process, as each citizen panel produced a Declaration regarding an improved management of the coastal water resources, and a better integration of science, policy, and citizens. The local workshops were complemented by a set of local conferences, extending the discussion that had started at the workshops to the local policy audience, and to a wider public with interest in the topics discussed. These day-long events were also part of the evaluation process.

This work at the regional level fed into a second European level workshop, where experiences were exchanged between not only the citizen panels but also the scientific project partners of each case study area. Over a 2,5 day event a Common European Citizen Declaration was developed. This document – presenting a set of recommendations regarding sustainable coastal water management and improved connectivity of actors – was presented at a day-long final European Public Conference, serving as a "public sphere" for cross-cutting dialogue between project partners, the transnational panel of citizens, European policy-makers, science-policy interface experts, invited speakers, and a wider public. Both of these events were also taken into account during the evaluation process. Figure 9.2 below describes this process visually, noting that the process starts at the European level, on the left side, moves to the local level, and back to the European level. It should be noted that the local Declarations are produced first, and then are followed by a common European Declaration.

9.2.2 Scope and objectives of evaluation

Evaluation is usually associated with the end of a process as a result of the formal accountability criteria applying to public budgetary lines. But in reality it can be launched at any stage of the policy process. There are three standard forms of evaluation: *ex-ante* evaluation carried out prior to the implementation of a policy as an input into its design – at EU level this is also known as regulatory impact assessment; *ex-post* evaluation is carried out at the end of the policy process and is meant to provide insight into how well the policy was implemented and with what results and/or effects; finally, *monitoring* delineates that form of evaluation that accompanies the policy process as a form of technical assistance. It is this latter type of evaluation that comes closest to describing the scope of evaluation within the AWARE project. It is also important to keep in mind that the object of evaluation in AWARE was not a policy per se but rather a process instituted in the framework of a research project, an equally relevant terminology and scholarship is that of 'action research', as used by Reason and Bradbury (2008). The evaluation methodology was drawn up by Giorgi and Pohoryles (2010) early in the project, before the first participatory events took place. From the point of view of the evaluation methodology it was useful to frame the process into singular phases:

(1) The *design phase* included the recruitment procedures and methods, the setup of the whole agenda, the ground rules for participation, the needed expertise and the information to be communicated to participants.
(2) The *development phase* encompassed all the activities performed during the process, the exchange of information and the application of participatory methodological principles during each event.

It is clear, however, that the design of activities is subject to iterative revisions: as the process developed, the application of the general methodological framework was refined and reviewed. This revision mechanism, although led by the managers of the process, was supported by the evaluation team by providing relevant information to the case study managers in order to achieve the target of the deliberative process.

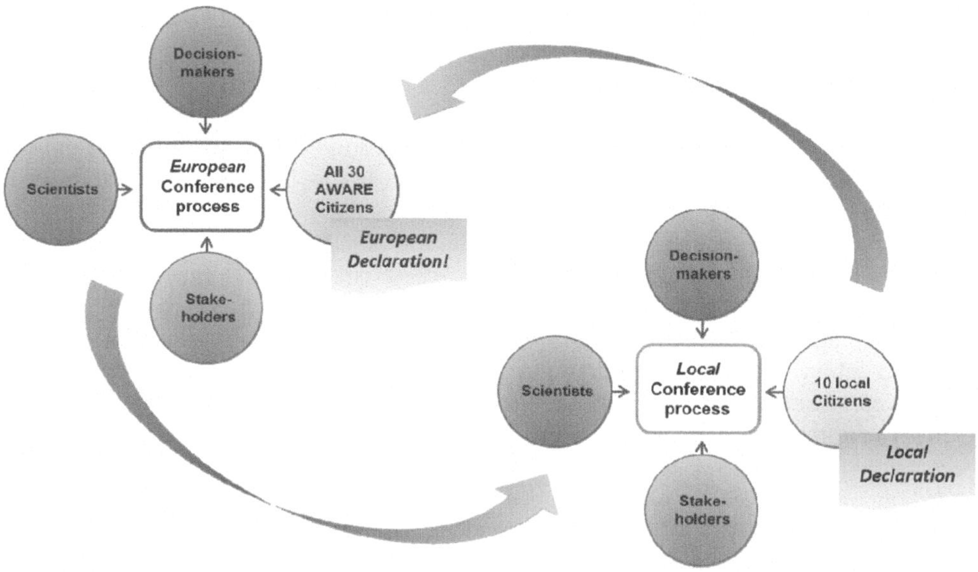

Figure 9.2 The AWARE knowledge brokerage methodology. A more detailed visual representation of the methodology is available in the AWARE Brochure "Bridging the science-policy gap: Best practices in citizens' participation", on the http://www.aware-eu.net/project website.

The objective of the AWARE evaluation was to describe and assess whether the choices made both during the design and preparation phase and the development phase:

- Affected (fostered or hindered) the deliberative process;
- Allowed for the sharing of knowledge among actors;
- Produced results, generated consensus and deliberations.

In this respect, the approach of the evaluation team, adopted during the consultation and participation events, included assessing how a particular method was used, with what results and attitudes, as well as the perceptions and positions communicated and adopted by the various actors along the process. The evaluation was shaped considering the particular contextual information depending on the three case studies and on the main object of deliberation, that is coastal water management.

Although ecological decision processes have the longest tradition in participatory approaches, still a formalisation of a common framework for evaluation methodology is lacking. Some principles that are at the basis of participation and widely accepted and applied (Renn *et al.* 1995) however, are relevant to be assessed by evaluation: fairness and competence. The first makes explicit the need for providing participants with equal opportunities to access and to act within the process. Fairness in this sense affects both the design and the development phases. Secondly, a process is competent when it considers the appropriate information and makes it usable by the actors involved through procedures and methodological choices.

Taking a more in-depth look into the evaluation methodology, Beierle (1999) identified four key components of any evaluation of a deliberative process: (1) representation; (2) the structure of the process or procedures; (3) the information used in the process; and (4) the outcomes and decisions arising from the process. All four criteria enclose into their definition the principles of fairness and competence. Table 9.1 is an adaptation of the structure provided by

Abelson *et al.* (2003) to the AWARE contents and scope, correlating each key component with some of the indicators used within the evaluation monitoring process. Indicators are in effect a list of questions to be answered by the evaluation team.

Table 9.1 Indicators within the evaluation monitoring process.

Representation	Procedures and rules	Information	Outcomes/decisions
Fairness and methods of the selection process	Credibility/legitimacy of process	Accessibility, readability, digestibility	Legitimacy and accountability of deliberation (responses to decision or input, more informed citizenry)
Characteristics of the sample (representativeness, education, sex, demographics, geography, etc.)	Relationships among stakeholders	Selection and presentation	Achievement of consensus over the decision
Characteristics of the experts (scientists, moderators, etc.)	The role of moderators and of evaluators	Methods for knowledge exchange	Participants' level of knowledge
Characteristics of the stakeholders (relationships, power, stakes, etc.)	Technical methods for participation	Local past experiences of participation-based policies	
Characteristics of the policy-makers	Degree of citizen control/input into agenda setting, establishing rules, selecting experts, information	Adequacy of time provided to consider, discuss and challenge the information	

Source: Adapted from Abelson *et al.* to fit AWARE scope, objectives and contents.

Each of the criteria listed in the table is briefly discussed below.

(1) Representation: this criterion addresses all the aspects taken into account for recruiting the participants to the process and how the recruiting activities were performed. Examples of indicators for this criterion are:
- How many citizens applied to participate in the consultation process? How many were finally selected?
- What stakeholders are represented? (Do these include all those with financial and/or knowledge resources able to influence or arrest the policy process?)
- Are citizens represented in the consultation process? What criteria were used for identifying participants? On what basis was this decision made?
- What recruitment methods were used? Which proved more useful and/or successful?

(2) The procedures and rules criterion evaluates the methodological choices made for framing the overall process and the singular events. The choices range over different aspects, from logistics of events to the enrolled moderator of the participatory session or the technique used for exchanging opinions among participants. Examples of indicators for this criterion are:
- How many meetings were planned and how many took place and when?
- What impact is the citizens' deliberation expected to have?
- Were experts involved? In what function and at what stage?
- Who played the role of moderator during the Local Workshop?

(3) The role of information criterion within the evaluation is to assess what and how information is selected, communicated and understood by participants.
- Which activities took place to communicate and exchange knowledge and information among participants?
- What type of knowledge (expert, tacit, local) was communicated or exchanged?
- How were the discussions structured?
- What methods were used to disseminate or exchange knowledge?
- What methods were used for problem-solving? (Were any ICT-based tools used to facilitate this process?)

(4) The final set of evaluation criteria considers the various potential outcomes of the public participation process. One aspect to be considered is whether the final deliberation has legitimacy and accountability for citizens themselves (consensus over the deliberation), policy-makers, stakeholders and public audience. The evaluation team has also to assess whether participation resulted in a satisfactory experience for the citizens and participants. Moreover, in AWARE, the specificity of themes led to the addition of some focused points to the evaluation scope, to be observed and assessed both by participants themselves and the evaluation team. Among these, the most preeminent was the measure of the participants' level of knowledge regarding three central topics: EU legislation, coastal environmental issues and the actual interface between science and policy. Examples of indicators are:
   - Did the knowledge of participants increase through the public participation process?
   - Did the attitudes of the participants change?
   - What were the citizens' expectation towards the process and how did they change?
   - Did policy-makers consider the citizens' Declaration worthwhile? What impressions did they have?

In summary, the evaluation addressed two macro-phases of the whole process: design and implementation. The design phase occurred not only at the beginning of the entire process for framing purposes, but prior to each participatory event. Moreover, since one of the evaluation purposes was to monitor and support the process, the phases were iterative: revisions and re-designs occurred during the development of the process to adjust it and fix eventual biases. The evaluation was inspired by two key-principles – fairness and competence – and used four basic criteria, inspired by the principles. For each criterion, a list of indicators was provided to help the evaluation team assess the methodological frame, the development of the participatory activities and participants' skills and attitudes.

9.2.3 Methods and rules for collecting information

The monitoring and evaluation of AWARE was undertaken over the course of 15 months; the AWARE evaluation team (partners in charge of the evaluation processes) developed the questionnaires for selecting participants as well as the questionnaires used to assess changes in knowledge, attitudes and skills among participants at the end of the knowledge brokerage processes. Each of the three case studies, as well as the European level events, was monitored by at least one project partner.

In order to collect all the relevant information for the evaluation, various activities and techniques were adopted. First, a collection of some factual information about the process was made, consulting the three case-study description and generally relying on the contextual information supplied by the partners directly involved in the organisation of the public participation process. This allowed for a global picture to emerge of the state of each case study, the geographic, demographic and cultural features, the stakeholder network present, and the policy-making levels and responsibilities.

The second element centred on observation: each citizen workshop and the corresponding preparatory meetings (both at European and local levels) were observed by the assigned evaluation team in order to document the precise nature of interactions as well as content and organisational issues. Participant observation helped to counteract any bias emerging out of the procedures. Figure 9.3a below gives a glimpse of one of the local workshops and Figure 9.3b highlights the final participatory moment of the project.

(a) (b)

Figure 9.3(a) Members of the Gulf of Riga citizen panel discussing local and regional challenges in coastal water management, in Jurmala, Latvia. (b) A selected panel of citizens presenting the Common European Citizen Declaration during the European Conference in Brussels, Belgium.

Third, informal interviews with citizens were undertaken before and after the workshops. The purpose of the interviews was to find out what and how much the citizen had learned during the workshop and how they perceived the interactions with the experts, with the AWARE project team and among each other. Interviews were also held in between the citizens' participation events with local and European policy-makers. At each point the information gathered from the participatory events were used as a basis for the interviews, and the lessons learned from these discussions were used as part of the iterative revisions described in Section 9.2.2.

The fourth and final element was a questionnaire completed by all participants after each workshop. This questionnaire consisted mostly of multiple choice questions but also included a few open questions. A related questionnaire was also disseminated to a range of networks of water managers, scientists, and policy-makers in the local case study regions and at the European level.

As an additional aid to the project and to its evaluation, a Policy and Stakeholder Advisory Group (PSAG) was established early in the project, whose members included representatives from the three case studies and from the European Commission. This body met three times during the course of the project, in conjunction with significant participatory processes, such as the first European Citizen workshop, the European Conference, and a final evaluation workshop. Indeed, the key policy-makers and stakeholders (9 were initially selected from the three case studies areas but in the course of the project the group gained further members) provided advice on the AWARE approaches and challenges; evaluated the process; and fostered the cross-sectoral communication in the case study regions. Several members were also highly involved in the local knowledge brokerage events, taking part in the local deliberations with the citizen panels. This group was in a position to highlight issues of process and outcome along the project's lifetime, at both the European and local levels, and their advice and input was taken into considerations in the iterative revisions mentioned above.

The evaluation team compiled the data and documentation necessary for carrying out the evaluation based on information provided by the local partners in charge of each case study, as well as through fieldwork observation of the public participation meetings in each case. Detailed reports were submitted for each case study and were used to compile a comparative analysis of the parallel processes (see Comardicea *et al.* 2011a and 2011b).

The final evaluation analysis combined the qualitative data of the observation documents, the interviews and the factual information, and the quantitative data derived from the questionnaires. The elements of the analysis mentioned above are discussed in the results section below, with the exception of the third element addressing informal interviews with the participants; the information gathered from those interviews were combined with the rest of the analysis and are presented throughout.

9.3 RESULTS AND DISCUSSION

9.3.1 Factual information about the process

The AWARE project focused on applying an integrated adaptive process as an innovative mechanism of knowledge brokerage and organisational learning to link research to policy development, in three pilot case studies of water systems in coastal areas. It should be reminded that the aim of AWARE was not to produce primary research activities, but only assess the existing evidence with the involved actors – water managers, the public, water management stakeholders, and scientists. Overall the knowledge brokerage process designed and implemented for AWARE proved to be well planned and structured, and largely carried out in line with the original plans. This is seen especially in the local case studies where most organisational and procedural aspects were evaluated positively by both the participants/citizens and the team of observers. For instance 80% of the North Sea case participants were "satisfied" with the overall process, and 70% were either "satisfied" or "very satisfied" with the discussion parts of the workshop; similar results were observed for the other two local workshops.

Within this positive assessment of the local proceedings, however, two distinct weaknesses could be highlighted. The first one concerns the interaction between the citizens and the scientific experts. All experts were asked to make their presentations during the local and European workshops as understandable for a lay audience as possible and with a few exceptions this part of the interaction was successful. What seemed most difficult was for experts to provide a complete set of information to participants in the short time of a workshop slot. The evaluation teams noted that in some cases even small bits of incomplete information surfaced later in the process in the form of erroneous assumptions on the part of certain citizen groups. This situation however, would be difficult to overcome in similar participatory processes, because it is impossible to predict which path the citizen deliberations will take and certainly it is not desirable to pre-determine this path. One possible solution might be to involve all available experts throughout the entire process to provide a *knowledge repertoire*.

The second weakness – especially at the local level – concerns the engagement of policy-makers, who were often difficult to involve in the AWARE activities. In those cases where they participated, the interaction with citizens remained on the level of political statements rather than taking the shape of an exchange of ideas, although this appears to be a systemic problem rather than an organisational weakness of this specific project (e.g. Mostert *et al.* 2007). Mostert and colleagues (2007) argue that the existing governance style is not used in multi-party approaches and, thus, participation with water management authorities often remains limited. Unlike science, the realm of policy-making is not concerned with understanding natural or social phenomena but rather with representing and/or weighting different political and economic interests in the region or a particular case. The active involvement of policy-makers in the process immediately raises issues on the legitimacy of any eventual decision taken on the bases of the deliberations formulated by a random group of ten citizens in each of the three cases. However, an alternative way to enable a more productive interaction between policy-makers, scientists, and citizens would be to engender a continuous informal discussion process between these actors, on key sustainability issues. Such an awareness-raising process could help bridge the gap between the citizens' lay knowledge and on-the-ground understanding of the topic, the understading of elected representatives, and that of research and social scientists engaged in the issue.

Indeed, this understanding was strengthened during PSAG meetings, when members (described in Section 9.2.3 above) who are local decision-makers mentioned the difficulty they are placed in when having to choose a path based on many opinions, uncertain scientific data, and a variety of lobbying groups. However, policy-makers at the European level – including the European member to the PSAG – were interested in the AWARE process due to the future implementation requirements – to include both scientific information and public participation – of the Marine Strategy Framework Directive. This interest was shared by most of the regional water managers as related to the implementation of the Water Framework Directive.

For the recruitment of stakeholders it was useful to separate them into groups depending on their level of influence and interest, as in the Gulf of Riga case study (Stalnacke *et al.* 2009). While taking into account those actors with high levels of interest and influence, AWARE also saw those stakeholders with high interest but low levels of influence as important and potentially in need of empowerment. Also, those stakeholders with high levels of influence but low interest were seen as playing a role in the context of decision-making. Interestingly, the scientific community was considered among those stakeholders that had both a low level of interest as of influence. It appears that until now the role of scientists in water management planning has been rather passive. Although they contribute knowledge, they are rarely more actively involved and engaged in the decision-making process. Recent trends however, show that research projects are starting to include sharing data and proposals targeted for policy measures, which address the science-policy gap. It is also interesting to note that in AWARE the initiative for engaging the stakeholders was put in the hands of the scientific community, which gave those invited increased credibility and incentive to participate. The stakeholder group most visibly missing from the table were representatives of industry. This was noted by all those involved in the project – including citizens, project partners, and interviewed experts at local and European levels. This lack of representation in fact occurred at both local and European levels. The approach of AWARE may indeed not have been useful for this group, and during the expert interviews that were part of the evaluation process it was mentioned that the industry group may have benefited from bilateral consultations prior and during similar participatory exercises.

The recruitment process for the citizen panels balanced the challenges of building a sufficiently differentiated panel of citizens, to represent the different typologies of users, and finding individuals sufficiently versed in English as to successfully participate in a pan-European group, with the identification of individuals interested in coastal water management and in taking part in an almost 2-year participatory process. Here the three case studies had rather different experiences. The Sacca di Goro case study, in particular, had difficulty meeting the English skills requirement, a fact which was compounded by the small size of the area under consideration. The resulting panel was in fact not fully representative of the Sacca di Goro social structure, since about half of the recruited persons hold a degree, while most of Goro citizens do not; similarly, only one fisherman was selected although most of the population in Goro is employed in fishing. This situation was balanced out by the stakeholders that were invited to participate in the local workshop, who more aptly represented part of the social structure of Sacca di Goro, together with its conflicts and interests. While the Sacca di Goro and the North Sea case studies received 11 and 20 qualifying applications, respectively, the goal of attaining 100 applications was achieved only for the Gulf of Riga case study, with applications received from Latvia and Estonia. This could be due to the fact that the Baltic Environmental Forum (BEF), an environmental NGO with good networks and experience in mobilising the public, led the recruiting process, while in the other two case studies the scientific and social science project partners leading the process had more limited mobilisation experience and networks. The recruitment of stakeholders and citizens was also highly dependent on the publicity methods used. All case studies relied on conventional dissemination methods such as online announcements, displaying posters in relevant locations such as universities, libraries, and so on. The BEF however, also published the announcement on the biggest portal for job search and vacancies in Latvia and Estonia. This can be considered as one key aspect of the success in reaching such a wide group of people.

In terms of citizen participation, the knowledge brokerage methods used can be described as expert assisted deliberations, including both factual learning and actual deliberations. In the first part a wide range of experts presented the scientific knowledge needed for participants to take part actively in the 2-year process – including natural science knowledge about coastal waters, policy frameworks, and social science oriented knowledge on issues like governance and participation. The second part – the deliberation – was moderated by carefully selected project consortium members (i.e. independent facilitators). Using structured discussions of the most important issues, these deliberations helped to develop the Citizen Declarations that were the main project outcomes at the local level. Additional moderation training was provided prior to the local knowledge brokerage meetings, especially regarding the intense knowledge exchange with 10 to 15 external experts, in order to ensure that the local events followed a similar moderation structure, and that challenges could be addressed in a similar manner. The sessions were carefully documented and transcriptions were made available. The evaluation indicates that the satisfaction with the moderators was relatively high throughout, as in the North Sea case where 80% were "satisfied" and 20% "very satisfied", results that are also reflected by the other two case studies.

9.3.2 Observation

Throughout all steps of the AWARE participatory process, knowledge was drawn from, and provided by, various participants albeit in different formats and in different amounts: *expert* knowledge was given mainly by scientists, *tacit and local* knowledge mainly by stakeholders and local policy-makers, *local and experiential* knowledge mainly by the involved citizens.

Due to the relatively small number of citizens in each group and the professional organisation of the deliberations, a lot of attention was paid to the ideas and opinions of each individual, which was reflected in the fact that participants perceived the events to be very fair. This was true not only for the local case study workshops but also for the two European level workshops. This ability to give a voice even to those citizens who are not as outspoken or dominant as others is clearly shown in the observation protocols as well as in the subjective assessments of the participants themselves. It is also positively reflected in the satisfaction of the citizens with the outcomes, and with the process itself. The European workshops however, faced the additional challenge of integrating three closely knit groups from different countries. Not surprisingly, this worked reasonably well for the two groups from the North Sea and Baltic Sea case studies because they had an international cross-border composition already. The Italian group was more difficult to integrate, given their already existing connection around a small geographic area, their close-knit social ties, and a noticeable language limitation in individual cases. This challenge, which was partly compensated by a process designed to pay attention and encourage individual opinions, presents an important lesson for similar future projects, especially those addressing international/cross-border issues.

The communication and exchange of knowledge and information among participants was mostly limited to presentations and discussions during planned participatory events – the workshops and conferences at the local and European levels. For practical reasons the citizens selected from across borders were not asked to commit to a process that would entail a continuous flow of information during the entire duration of the project. However, the Sacca di Goro citizen panel, again representing a more compact geographic and social landscape, spontaneously organised themselves to keep in contact and continue working on project outputs in between the official project events. This continuous process resulted in a higher case study group cohesion and ownership in the project, but not necessarily a significantly higher level of satisfaction with the output. In the Sacca di Goro case almost 60% of the citizens were "very satisfied" but over 30% were "satisfied on average". This compares with the other two case studies: in the North Sea case 50% of the citizens were "satisfied" with the outcome – the set of recommendations produced in the local workshop – and 20% were "very little satisfied". In the Gulf of Riga 50% of participants were "very satisfied", but again 20% were only "satisfied on average"; in this particular case study the challenge of producing the Citizen Declaration was compounded by the fact that for practical reasons the local conference (where the Declaration was publicly presented) was scheduled only two days after the local workshop where the Declaration was developed by the citizen panel. However, future similar projects could use innovative means of supporting such a sustained information exchange even across border, using video conferencing for instance (standard electronic means of communications like e-mail or discussion forums are shown to not be very successful, and AWARE's trial of a Facebook page also met with low resonance among participants).

Even with minor differences in the assessment of the three local Declarations described above, by and large the citizens were satisfied with how their views were heard, processed and eventually integrated into the final policy-relevant documents – the local Declarations. However, a significant amount (between 20–30%) of citizens achieved only an average level of satisfaction, for which the main factor was time pressure. For instance one participant from the Gulf of Riga stated that the Declaration statements were too broad and "too standard" in the end. In the local as well as European level deliberations around the Declarations, a common lesson learned was regarding the efforts to make these outputs fit a "policy audience" – this was attempted by the project partners given the final audience planned as recipients of the Declaration, but it was resisted in many cases by the participants, as the language was perceived to lose the desired expressions. Expert opinions regarding the Declarations however – policy-makers and other stakeholders, although acknowledging the novel perspectives they brought to the problems at hand, in many cases also saw the outputs as *naïve* and lacking practical experience. This reflects perhaps the observation made at the beginning of this section that knowledge was given and absorbed by all those involved in the knowledge brokerage process, but was presented and perceived in different ways. Also, it harkens back to comments made throughout meetings with the PSAG members and throughout expert interviews, that the difference in "languages" or expression styles between the science, policy, and public realms, still have a long way to go before common understanding is achieved.

The way in which the common European Citizen Declaration was developed, using information from all three local Declarations, was generally seen more critically than the drafting of the local recommendations. Even though the citizens were of the opinion that their personal views were reasonably well reflected in the European Declaration – again reflecting the fairness of the process – they did not feel equally strongly that the ideas and spirit of their local Declaration had been translated into the European document. This is partly due to the fact that the more difficult process of working with a pan-European group of 30 citizens to combine three distinct sets of recommendations into one was planned for one 2,5 day workshop, which may not have given sufficient time for the work. In addition, the expectations for the common European Declaration were rather high, given their target of EU policy-makers, which increased the pressure of using policy-oriented language and may have altered in some cases the spirit of the original local Declarations.

9.3.3 Questionnaires

The evaluation of the participants' knowledge, skills and attitudes at the first European level workshop showed that gaining *knowledge for action* was the most frequently mentioned expectation, that is learning about the problems faced by coastal management and learning what can be done about it, collectively and also as individuals. This finding reflects the selection criteria of the citizen panels to include members who are interested in the topic; in addition, a few participants were motivated also by professional interests, as their work involved either working with citizens and/or working on coastal management. Regarding the level of knowledge with which citizens entered the process, the baseline assessment done at the first European workshop measured relatively high levels: 70% reported being knowledgeable about their coastal area (either in learning about it in preparation of the meeting or having knowledge in advance); 56% had a pretty good knowledge about the Water Framework Directive or having learned about it in preparation of the meeting; and 30% knew about the science-policy interface or had learned about it in preparation of the meeting. The learning outcomes from this first workshop were measured by asking the participants to assess the added value they had received from the meeting. Only two participants reported learning only a little and the overwhelming majority, or 93%, reported learning a great deal. The learning aspect was also often mentioned in the comments made by participants to the open questions.

Questions relating to the quality of the invited speakers were positively evaluated by the citizens, with a higher level of satisfaction regarding speakers than regarding the workshop debates and discussions. On the one hand this shows that in the process of designing the workshop agenda care was taken to acknowledge the wishes of the citizen panels for the types of stakeholders and topics to be discussed; on the other hand however, this is also a reflection of limited time and a focus on imparting knowledge and having a good basis for discussion, which necessarily limits the time available for the intended discussions. Additionally, as mentioned in Section 9.3.1 above, the involvement of policy-makers in the knowledge brokerage events was relatively difficult and the interaction was not perceived as an exchange, but rather as a political statement, not committed to acknowledge the citizens' point of view and discussion outcomes. As an example in the Gulf of Riga workshop evaluation, even though most sessions were rated with a high degree of satisfaction, the only session over the two days that received a score of "not very satisfied" was the one related to policy solutions – which may reflect the difficulty of understanding a complicated presentation, not tailored for the audience.

Regarding the assessment of the amount of knowledge gained in the local workshops, in addition to the European workshops, the local questionnaires reflected similar results as those from the first European workshop. In the North Sea case for instance 60% of the citizens had learned about the Water Framework Directive in preparation for the meeting, or had received the knowledge from the European workshop, and 30% had a pretty good knowledge beforehand. Regarding local issues of environmental pollution, and the topic of the science-policy interface, 50% learned about them in preparation for the meeting and 30 or 20% respectively, had a pretty good knowledge beforehand. From the comments given in the open questions of the questionnaire it was also noted that many citizens had wished for more time within the participatory events to exchange, and to "digest" the knowledge gained. Indeed, an even more direct participation was desired by the citizens such as by strengthening the direct interaction with those presenting, and giving them the opportunity to report at the end of the day. Citizens also consistently enjoyed the glimpse into the various ways scientists, policy-makers, and the public can and do cooperate, and an extended opportunity to exchange views might in future projects also aid in bridging the "language" gap between the different actors, which was mentioned in Section 9.3.2 above.

In the Gulf of Riga there were some interesting differences noted between the same questions asked to the stakeholders and the citizens. While stakeholders generally stated to have a pretty good knowledge, especially about the local coastal environmental problems and the Water Framework Directive, the majority of citizens had learned about these issues in preparation for the workshop. In addition, over 60% of the stakeholders that responded stated they have "learned a little" from the workshop, with only over 30% having "learned a great deal". In comparison, 90% of the citizens stated they "learned a great deal" while only 10% "learned a little". This result also mirrors several comments made by the citizens during the workshop, which describe themselves as "only lay citizens", thus potentially underestimating their existing knowledge or capacity to contribute expert knowledge to the proceedings.

In the Sacca di Goro case study the results are fairly similar to the Gulf of Riga, where stakeholders generally have a pretty good knowledge especially regarding local coastal environmental issues, but citizens describing lower levels of knowledge, or stating they had learned about the issues in preparation for the workshop. A similar comparison between the amounts of learning occurring among stakeholders and citizens is seen here: while 37% of stakeholders "learned a great deal" and 50% "learned a little", 78% of citizens "learned a great deal" and only 22% "learned a little". Overall, stakeholders held a greater knowledge of the topics than citizens, and did not learn about them in preparation for the workshop. Perhaps reflecting a lower level of confidence about expert knowledge – in a participatory activity that includes many actors viewed by citizens as experts, such as scientists, stakeholders, and policy-makers – the citizens responded that they "have just heard about" some topics, even though they had already gained significant knowledge about these topics at the prior European level workshop. It should also be remembered that citizens were selected based on interest in the topic, not on prior knowledge, so the results show a natural acquisition of knowledge and the process of becoming comfortable with it. The fact that so many citizens across the different case studies state their knowledge as being the result of preparation for the meetings, however, may also reflect a greater commitment on their part, as long-term participants in the project, as well as an outcome of capacity being built through AWARE's knowledge brokerage activities in which citizens develop ownership and gain confidence to share and improve on their knowledge, and provide valuable recommendations.

Lastly, questions regarding the process of the workshops – including logistics – were positively evaluated in all the case studies. In the case of the Gulf of Riga for instance, the catering was most appreciated, followed by the location and the overall organisation of the workshop. The location in fact played an important role – as Jurmala was chosen due to its location on the Gulf coast – and this was appreciated by all participants, including the invited speakers. In the Sacca di Goro case the catering was also the logistical aspect most appreciated – with 80% scoring it at "very much satisfied", followed by location and organisation. In addition to the local events, the participants at the first European workshop were also asked to evaluate the organisation of the meetings, and mentioned the following challenges: too much information packed in each day, too little time overall, too few social events and opportunities to network and exchange knowledge with each other. Although the participants were compensated for their time, and although the events were scheduled to fall across both working days and weekend days, some participants suggested holding the meetings only on weekends so as not to lose working time.

9.4 CONCLUSIONS AND LESSONS LEARNED IN THE EVALUATION PROCESS

One of the main lessons from the AWARE knowledge brokerage process was learned as a consequence of the remarkable commitment shown by the citizens selected in the panels: their mandate was ambitious and managing their expectations within this experimental participatory exercise was a challenge. At the same time it was noted that a high level of commitment and team cohesion was reached within the project and that in fact the citizen participants delivered a high quality set of outputs, providing efforts beyond what had been planned within AWARE. This bodes well for similar participatory processes, and highlights the importance of future citizens-science programmes, as described in Chapter 10. Policy-makers – who joined the process primarily as part of the local conferences, the PSAG, and through online surveys – showed a range of different feelings towards the AWARE scope and procedures, ranging from enthusiasm to coolness. Stakeholders who were present actively joined the process and showed their own points of view and interests, as highlighted in Section 9.3 above.

Some final considerations can be drawn, most of which have already been explored above, and which are based on several sources: the observations and notes of the case study evaluations; informal discussions between all AWARE partners around the various events; online surveys with local and European policy-makers, scientists, and stakeholders; interviews at the local and European levels; the questionnaires gathered after workshops and conferences and highlighted in Section 9.3.3 above; and the final AWARE Evaluation Workshop, to which the PSAG members were invited, along with a number of external science-policy interface experts, and which is described in more detail in Chapter 10. A leaner version of these recommendations, aimed at water managers and policy-makers, but relevant for a wide range of participatory processes, were included in AWARE's best practices brochure. The eight recommendations, listed below in detail, focus primarily on three areas: a) bringing as much as possible *all* involved stakeholders together, across borders but with a local implementation emphasis, to interact regularly; b) engaging with well-connected NGOs and with an interested core body of scientists throughout the length of any such process in order to improve environmental awareness and the transference of scientific knowledge; and c) sharing the lessons learned from such processes among other managers, decision-makers, and the public at all levels from local to pan-European (Comardicea *et al.* 2011c).

First: *Bring citizens, scientists, stakeholders, and decision-makers together in a participatory knowledge brokerage process to improve decision-making at the local level and increase ownership of challenges affecting citizens across Europe.* Implementing a knowledge brokerage process for improving decision-making at the local level can help increase ownership: citizens and stakeholders have a higher incentive to participate around local issues where they will be able to track and measure the impact of the decision process. Nevertheless, working at larger, European, scales provides more relevance to the process for all actors involved, as it also includes the top-level policy framework for sustainability issues. Thus, depending on the objective of each participatory process, a compromise between these two perspectives – local/national and supranational – should be found. At all scales ownership and personal involvement is significantly increased by clearly defining the outcome that is expected from the knowledge brokerage process. In fact, as the AWARE local workshops showed, maintaining the sense of ownership over the outcomes (the content and wording of the Citizen Declarations), if not the entire process, is crucial for involved citizens. A loss of ownership – happening through either poorly managed expectations, insufficient preparation and discussion time, or excessive "tailoring" of citizen outputs to a particular audience – decreases the engagement of citizen participants in the process, defeating thus the hoped for outcomes.

Second: *Focus on local level implementation while also feeding the lessons learned back into European level policies.* A knowledge brokerage process may best influence the local implementation process of European Directives, since these provide some flexibility and room to manoeuvre at the national and local scales. The participatory process also benefits significantly from informal-formal regular opportunities for various actors to interact, which is easiest at the local levels. However, it is also important for local participatory processes to provide feedback on the ways in which European Directives can be implemented, which will be useful at the EU level in the design of future policies.

Third: *Take into account cultural and empowerment differences when implementing a knowledge brokerage process, especially to ensure a fair level of trust in the process outcomes.* Taking into account cultural and empowerment differences is important before implementing a knowledge brokerage process, especially regarding trust in the outcomes of such a process: it is important that the broader public – from among which a representative panel will be selected – believes that concrete results can be expected from the process, given clearly defined outcome goals. These outcomes are not necessarily a specific set of decisions, which are ultimately to be taken by legitimate powers at EU, national, regional or local level, but may be more broad and informal outcomes (actions or initiatives) that can help to determine policy strategies supported by a deeper consensus, built upon a greater awareness of the issues at stake from all actors concerned. Cultural aspects also impact on all manner of participatory activities – such as level of input given, participation for instance in online forums after the life of a project, trust in new initiatives, and so on.

Additionally, proficiency in a common language – English in the case of AWARE – matters in supranational participatory processes since it can have direct positive or negative consequences on outcomes. The Gulf of Riga group for instance was the only one of the three case studies that participated in the local events without speaking their native language. This decision was made due to efficiency issues as part of the initial project design. Although this did not seem to affect the proceedings for most participants, some citizens may have been more comfortable expressing their opinions in their own language. In the Sacca di Goro case however, Italian was used for the participatory process as the area was not transboundary. Within the European-level events however, English was used and the Sacca di Goro citizens showed a comparatively lower proficiency in English. Language considerations are thus crucial to the design and implementation of participatory processes.

Fourth: *Bring citizens, scientists, stakeholders, and decision-makers together regularly to allow for trust building and effective learning and knowledge sharing.* Implementing a knowledge brokerage processes in a geographically delimited area is more efficient: the necessary permanent or regular informal processes involving different types of actors to support decision-making will be more effectively carried out. Such regular opportunities for interaction – focused in time and space on a particular area and topic – also allow the activities to take place in a common language. This speaks also to the balance that should be sought between the local and the pan-European aspects of citizens-science projects, as transnational aspects such as using a different common language and exploring culturally-specific aspects of the same societal challenges are relevant at the broader scale.

Fifth: *Involve local civil society actors to best reach and engage with the broad public.* Involving local NGOs and social actors is important to reach and engage a broad public: because of their good knowledge of local societal structures, they are able to act as "mediators" between actors. Well-connected and trusted NGOs can also act as catalysers of action and as "multipliers" to ensure a wide dissemination that reaches all types of lay citizens – not only those with access to most resources and networks, or those groups of selected individuals that consider themselves especially environmentally-friendly. In addition, NGOs versed in facilitating groups and public events could also help to moderate the participatory events. The AWARE workshops and conferences showed that good facilitation can be crucial in drawing out the right discussion points from a variety of speakers and in engaging with the participants. It was also clear from the evaluations of these events (through the questionnaires) that participants highly appreciated good moderation of the discussions.

Sixth: *Engage all relevant actors to the extent possible, particularly those from relevant industry or economic sectors.* Try to reach actors from relevant industry or economic sectors by engaging on concrete topics with industry representatives and federations who are likely to arrange regular contacts between industry actors, policy-makers and science. For instance, representatives from ports, harbours, and the industry were lacking in AWARE; as discussions often turned toward pollution in the Gulf of Riga events, some questions pertaining to these industry actors remained unanswered, thus decreasing the quality of the deliberations and of the citizen-developed outputs. Future EU funded knowledge brokerage processes between scientists, citizens and policy-makers should aim to connect with the European Innovation Partnership on a range of sustainability challenges, for instance water efficiency.

Seventh: *Involve a team of scientists throughout the knowledge brokerage process. "Scientific ambassadors" could communicate critical information to citizens, business representatives and policy-makers alike.* Different actors seem at times to almost speak a different language. Even more, as already noted by Ostrom (2009), the differences in concepts and languages amongst scientific disciplines may already limit our understanding of the natural and socio-economic processes. This perception was heard in informal conversations throughout various events when describing some panel discussions or a presentation particularly difficult to understand (see Section 9.3.3 above). Much of the discussion the citizens had regarding improving connectivity between scientists and policy-makers in fact, had to do with trying to bridge this gap of understanding and "speak the same language". Thus, AWARE has shown that involving a team of scientists for each piloted case study throughout the knowledge brokerage process is significant: scientific knowledge should be shared with a broader public, and the academic expert teams should accompany the whole knowledge brokerage process, also in designing the output of the process with the citizens' panel, to ensure that relevant knowledge is accessed, understood, and processed accurately. To help with complex sustainability issues, *"scientific ambassadors"* could "translate" critical information for citizens, business representatives and politicians alike. These experts should have cutting edge knowledge of research advancements in a given sustainability domain, personal communication skills, a mind open to a broader dialogue, and an understanding of the socio-economic implications of their research. Increasing academic engagement with projects similar to AWARE is one way to do this. However, the scientific award system could, in addition to a scientific career, recognise and encourage individual participation in this kind of participatory initiatives, as well as efforts done to tailor scientific results to wider target audiences.

Eight: *Exchange best practices and share the lessons learned from already implemented knowledge brokerage activities with others engaging in such activities at local or regional levels.* To increase the relevance and appeal of participating in more innovative knowledge brokerage processes, existing processes also need to be identified and reviewed: exchanging and comparing results is important for action learning and for improving the visibility of knowledge brokerage activities at local and regional levels across Europe. Feedback on this kind of experience is also needed at the EU level, where active involvement of different parties through participatory processes is encouraged, particularly on sustainability issues.

REFERENCES

Abelson J., Forest P. G., Eyles J., Smith P., Martin E. and Gauvin F. P. (2003). Deliberations about deliberative methods: Issues in the design and evaluation of public participation processes. *Social Science Medicine*, **57**(2), 239–251.

Alemanno A. (2009). The better regulation initiative at the judicial gate: A trojan horse within the commission's walls or the way forward? *European Law Journal*, **15**(3), 382–401.

Allan C. and Stankey G. (2009). Synthesis of lessons. In: Adaptive Environmental Management – A Practitioner'S Guide, C. Allan and G. Stankey (eds.), Springer, Jointly published with CSIRO Publishing, Collingwood, Australia, pp. 341–346.

Beierle T. C. (1999). Using social goals to evaluate public participation in environmental decisions. *Policy Studies Review*, **16**(3–4), 75–103.

Biswas A. K. (2004). Integrated water resources management: A reassessment. *Water International*, **29**(2), 248–256.

Borowski I. and Pahl-Wostl C. (2008). Where can social learning be improved in international river basin management in Europe? *European Environment*, **18**(4), 216–227.

Castillo A., Torres A., Velázquez A. and Bocco G. (2005). The use of ecological science by rural producers: A case study in Mexico. *Ecological Applications*, **15**(2), 745–756.

Comardicea I., Luè A., Pohoryles R., Schmidt M. and Tandon A. (2011a). Comparative Report On Case Studies. AWARE Project Deliverable D 3.2, available for download at http://www.aware-eu.net/ (accessed 10 April 2012).

Comardicea I., Horvath L., Luè A. and Schmidt M. (2011b). Comparative Report On Learning Outcome Evaluation. AWARE Project Deliverable D 3.3, available for download at http://www.aware-eu.net1/ (accessed 10 April 2012).

Comardicea I., Semmling E. and Sessa C. (2011c). Bridging the Science-Policy Gap: Best Practices in Citizens' Participation. AWARE Project Deliverable D 4.3, available for download at http://www.aware-eu.net/ (accessed 10 April 2012).

Endter-Wada J., Blahna D., Krannich R. and Brunson M. (1998). A framework for understanding social science contributions to ecosystem management. *Ecological Applications*, **8**(3), 891–904.

Folke C., Hahn T., Olsson P. and Norberg J. (2005). Adaptive governance of social-ecological systems. *Annual Review of Environment and Resources*, **30**, 441–473.

Giorgi L. and Pohoryles R. (2010). Evaluation Methodology. AWARE Project Deliverable D3.1, available for download at http://www.aware-eu.net/ (accessed 10 April 2012).

Gregory R., Ohlson D. and Arvai J. (2006). Deconstructing adaptive management: Criteria for applications to environmental management. *Ecological Applications*, **16**(6), 2411–2425.

Holling C. S. (1978). Adaptive Environmental Assessment and Management. John Wiley, New York.

Huitema D., Mostert E., Egas W., Moellenkamp S., Pahl-Wostl C. and Yalcin R. (2009). Adaptive water governance: Assessing the institutional prescriptions of adaptive (co-)management from a governance perspective and defining a research agenda. *Ecology and Society*, **14**(1), 26.

Kusel J., Doak S. C., Carpenter S. and Sturtevant V. E. (1996). The role of the public in adaptive ecosystem management. In: Sierra Nevada Ecosystem Project, Final Report to Congress, Vol. II, Assessments and Scientific Basis for Management Options, University of California, Centers for Water and Wildland Resources, Davis, CA, pp. 611–624.

Leach W. D. and Pelkey N. W. (2001). Making watershed partnerships work: A review of the empirical literature. *Journal of Water Resources Planning and Management*, **127**, 378–385.

Mostert E., Pahl-Wostl C., Rees Y., Searle B., Tàbara D. and Tippett J. (2007). Social learning in european river-basin management: barriers and fostering mechanisms from 10 river basins. *Ecology and Society*, **12**(1), 19.

Newig J., Pahl-Wostl C. and Sigel K. (2005). The role of public participation in managing uncertainty in the implementation of the water framework directive. *European Environment*, **15**(6), 333–343.

Norton B. G. (1998). Improving Ecological communication: The role of ecologists in environmental policy formation. *Ecological Applications*, **8**(2), 350–364.

Norton B. G. (2005). Sustainability: A Philosophy of Adaptive Ecosystem Management. University of Chicago Press, Chicago

NN (ND). Processos Participatius I Directiva Marc De l'Aigua. Spanish Only, Available Online At http://aca-web.gencat.cat/aca/documents/ca/participacio/portal/index.pdf (accessed 10 April 2012).

Ostrom E. (2009). A general framework for analyzing sustainability of social-ecological systems. *Science*, **325**, 419–422.

Pahl-Wostl C. (2007). Transitions towards adaptive management of water facing climate and global change. *Water Resources Management*, **21**, 49–62.

Reason P. and Bradbury H. (eds.). (2008). 2nd edn, Sage Handbook of Action Research: Participative Inquiry and Practice. Sage Publications, London.

Renn O., Webler T. and Wiedelmann P. (eds.) (1995). Fairness and Competence In: Citizen Participation: Evaluating Models For Environmental Discourse, Kluwer Academic Press, Boston, MA, pp. 35–86.

Roux D. J., Rogers K. H., Biggs H., Ashton P. J. and Sergeant A. (2006). Bridging the science-management divide: Moving from unidirectional knowledge transfer to knowledge interfacing and sharing. *Ecology and Society*, **11**(1), 4.

Stalnacke P., Bryhn A. C., Hakanson L., Nagothu U. S. and Veidemane K. (2009). The Gulf of Riga Case Study: Inception Report. AWARE Project Deliverable D1.2, available for download at http://www.aware-eu.net/ (accessed 10 April 2012).

Walters C. J. and Holling C. S. (1990). Large-scale management experiments and learning by doing. *Ecology*, **71**, 2060–2068.

Chapter 10

Enhancing connectivity between scientists, citizens and policy-makers in the water sector and beyond

Carlo Sessa[1], Camino Liquete[2], Francesca Somma[2], Irina Comardicea[3] and Elsa Semmling[3]

[1]AWARE Project Coordinator, Institute of Studies for the Integration of Systems, Via Flaminia 21, Rome, Italy
[2]Joint Research Centre – European Commission, Institute for Environment and Sustainability, Ispra, Italy
[3]adelphi, Berlin, Germany

10.1 INTRODUCTION: EX-POST EVALUATION OF THE AWARE POTENTIAL IMPACT

The goal of the AWARE project has been to demonstrate a new way of "connecting research, people and policy-makers in Europe to achieve sustainable water ecosystem management". The project used a variety of methods and activities to achieve this aim – including workshops, interactive conferences, online surveys and personal interviews.

Personal interviews with decision-makers in water management were conducted at the EU and international level (AWARE, 2011c) to assure a more personal approach and to get a more complete ex-post evaluation of the AWARE approach that could not be received via the online survey, as well as to capture further insights and recommendations from stakeholders that may not have participated in the AWARE workshops or conferences. The interviews at the EU and international level were conducted in the period between the AWARE EU Conference (June 2011) and the Ex-Post Evaluation Workshop for water managers (October 2011). The objectives of these interviews were to:

- Gain feedback on the perception of policy and decision-makers on active citizens' involvement in policy, and especially on their perception of the AWARE approach;
- Identify a range of reactions of policy and decision-makers to the results produced by AWARE, specifically to the citizens' recommendations for EU policy-makers;
- Produce recommendations on enhancing the connectivity between research and policy-making through participatory processes.

The institutions targeted by the interviews at the EU and international level included the European Commission, international and national research institutions, national ministries in charge of water and the environment, national water agencies, as well as European and international water organisations and platforms, such as the Euro-Mediterranean Information System on know-how in the Water sector (EMWIS), the European Federation of National Associations of Water and Wastewater Services (EUREAU), the European Water Partnership (EWP), the Water supply and sanitation Technology Platform (WssTP) and the EUREKA Cluster ACQUEAU.

A further and final step of ex-post evaluation was to convene a workshop inviting the members of the AWARE Policy and Science Advisory Group (PSAG) and a number of selected water managers to attend a full day meeting. The workshop was attended by representatives from ONEMA (France), the Latvian Ministry for Environment and Regional Development, the European Commission – DG Environment, the Catalan Water Agency (Spain), EMWIS/SEMIDE, MEEDDAT (France), the Coordination Centre for ICZM (Belgium), ISPRA (Italy), the Italian Ministry of Environment, the River Po Authority (Italy), the Estonian Ministry of Environment, EWP, and the Université de Versailles Saint-Quentin-en-Yvelines (France).

10.2 MAIN CONCLUSIONS ABOUT THE AWARE APPROACH AND ITS POTENTIAL USE TO ENHANCE THE SCIENCE-POLICY INTERFACE

The main conclusions drawn from the ex-post evaluation of the whole AWARE approach are as follows:

- All interviewees valued the AWARE experiment in a positive way, but highlighted the lack of involvement of the general public. Several of them inquired about the selection procedure and saw the small pool involved as a limitation/drawback of the process; many thought that the people involved represented an elite of environment-friendly intellectuals rather than lay citizens; the necessity was expressed at several stages during the interviews to involve larger masses of people.
- All the consulted experts and decision-makers welcomed the approach and the outcome of the AWARE process; many recommended the use of social networks or local scale dissemination actions to spread ideas and awareness and to foster participation by a larger public. At the EU level, it was suggested to disseminate the AWARE results in the framework of the 2012 Blueprint to safeguard Europe's water resources, in order to work on a "public participation charter". Based on these proposals the AWARE partners developed guidelines and examples of how to apply participatory processes in water management based on the AWARE experience (AWARE 2011b and this book).
- The involvement of industry stakeholders was an issue raised several times since they were mostly missing in the process. However it was recognised from the design of the AWARE process and from the interviewees' own experience that involving industry actors in such a process is in general not an easy task; for example SMEs cannot necessarily afford to spend time in participatory processes during their daily business. One suggestion was to try to take the AWARE results to federations in countries where these are likely to arrange regular contacts between industry actors, policy-makers and science (for example in Germany, the Netherlands or Belgium).
- In general, the experts and decision-makers consulted thought that the kind of participatory process piloted by AWARE would have little-to-no direct impact on coastal water quality and sustainability; some respondents conceded an indirect effect, or an effect only in case of larger-scale/long-term involvement of the public; others deemed a strong intervention (in the form of incentives, etc.) by local or national governments as necessary to achieve a significant impact.
- The AWARE approach was felt to be applicable to other policy areas. However the general feeling was that citizen involvement is more effective and efficient at the local level (within the everyday life horizon of the lay citizen) and, as the scale goes up to national or European level, the problem becomes more and more vague and impersonal, showing little pertinence to the citizens' personal experience. Such conclusions derived in large part from the past experience of interviewees in participatory processes, but was not shared by all of them. In a few cases, the interaction between the EU, national and local levels was seen as unavoidable, and the AWARE method was considered suitable to be replicated at all these levels to achieve an integrated view of universal sustainability issues.
- The reaction of the interviewees to the recommendations presented in the AWARE European Citizens' Declaration was unanimous in that, although positive, such recommendations are for the most part not well suited for the EU level. In fact, citizens did not seem to realize that European Directives that are currently under implementation (see Chapter 3 of this book) give little room for changes in their spirit or technical specifications and, rather, their influence could be stronger at the local level in defining managing strategies to achieve EU policy targets. Some of the citizens' recommendations are evaluated as very naïve and idealistic, typical of an educated (or scientists driven) pool of people, and difficult to implement at large scale: they would require a strong collective effort for public involvement, or top-down enforcement by national governments that affects people's behaviour by means of taxation, for instance.
- It was generally felt by the majority of experts and decision-makers consulted that there is a gap in the dialogue between scientists and policy-makers, which needs improvement; the dialogue is often unidirectional, with policy-makers asking the scientists for advice but not with scientists directing their research to answer policy-makers' questions. It was also felt that at EU level such dialogue works much better than at national or local levels, and the EU itself has its own body of researchers to rely on.
- It was frequently highlighted that the interface between policy-makers and the public is not as efficient as it should be. Scientific knowledge of lay citizens is generally low, thus hampering further the possibility of dialogue; the contact between local managers and the general public, however, improves as we move from the EU to the local level, as in the case of the River Basin Management Plans (RBMPs) where public involvement is a mandatory requirement explicitly outlined in the Water Framework Directive (WFD).
- As for having a significant impact at EU policy level, a fundamental requirement would be to institutionalise and make the application of the AWARE method more systematic. This would be useful especially to establishing trans-boundary activities and connections between policy-makers, scientists, stakeholders and citizens at the level of international river basin organisations. The AWARE approach however, needs endurance to be successful, because the best practises drawn from this experience need to be implemented at all levels, and acting on a single

project basis would be not enough (usually when a project is finished partners move to other subjects, there is no follow-up, no further funding and the method would be eventually forgotten). In this respect, the AWARE approach should be disseminated primarily to EU level institutions that hold a political representation of the European citizens, namely the European Parliament and the Council, as well as the national, regional and local governments. These institutions have competence to modify the legislation and open policies to citizens' participation. Another important EU institution that could promote the AWARE approach is the Committee of the Regions, which gives voice to sub-national authorities in the development and implementation of EU laws and can address the European Parliament directly.

10.3 FUTURE PROSPECT: HOW TO ACHIEVE A SNOWBALL EFFECT?

A key question that arose in different meetings and interviews as well as in the ex-post evaluation workshop was how to achieve a snowball effect, scaling-up the AWARE pilot project to produce a wider impact. Two directions of possible impact have been envisaged, as illustrated in Figure 10.1:

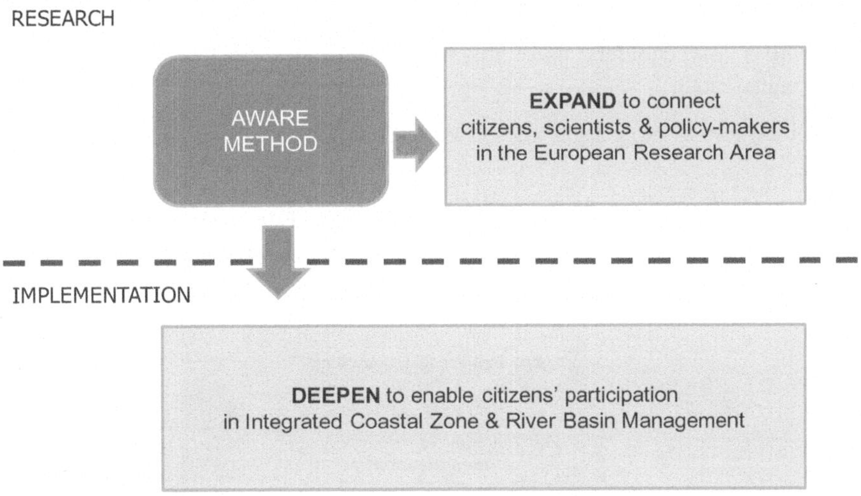

Figure 10.1 How can the AWARE methodology achieve a snowball effect?

- **Deepen** the AWARE approach in the water sector, in order to support the implementation of the Integrated Coastal Zone Management (ICZM) and the Water Framework Directive (WFD) across Europe, ensuring a more effective participation of citizens and stakeholders. The current situation of participation in the WFD context, and future prospects opened by new methods of participation like the one experienced in the AWARE pilot, were discussed in depth during the ex-post evaluation. It seems that water policy is still made "far away" from citizens, and although public participation is explicitly envisaged for instance in Art. 14 of the WFD, it is rarely implemented to the level of lay citizens; when it is implemented it is not consistent and feedback from the AWARE interviews, surveys, and involvement of the PSAG members repeatedly highlighted the lack of coordination, guidelines, and best practices in implementing such processes. There are participatory experiences at local or regional levels, but the recommendations stemming from these experiences rarely influence the decisions of politicians, which are taken unilaterally, often listening only to specific interest groups. In Catalonia, for instance, the process of formulation of the river basins management plans included a participatory process that engaged up to 20,000 participants (including lay citizens); but the ideas extracted from that participatory process were not taken into account for the modification of the plans in the Parliament by the politicians. By the same token, public participation is more and more important in the daily work at the Belgium ICZM Coordination Centre, joining stakeholders in a bottom-up process that develops scenarios at a small-scale and matures a vision to be addressed to the higher level (e.g. the Flemish region). However, the whole process engages stakeholders and water managers of the ICZM Centre and not policy-makers. The consulted experts and decision-makers agreed that the AWARE approach has the potential to improve the situation, provided that the right conditions to mainstreaming the approach are put in place – including a top-down intervention – as the process is too costly and complex to achieve the needed critical mass only by means of piecemeal and fragmented bottom-up initiatives.
- **Expand** the application of the AWARE method in other sectors, to other sustainability topics and related science-policy interface needs. Indeed, while AWARE addressed coastal water management specifically, the methodology could also be useful to deal with other sustainability and social challenges that unfold in the

long-term. AWARE-like knowledge brokerage processes can be particularly useful for issues that transcend electoral timelines, as they may contribute to building consensus-seeking processes, profiding for a more robust understanding of complex problems and a greater commitment towards durable actions and policies. Thus, future European projects inspired by the AWARE method could bring researchers and citizens from across Europe together in collaborative research experiences, to address cross-cutting societal and sustainability challenges that Europe is currently facing, including energy and climate change, and sustainable transport among others. The crucial point of such projects must be citizens' participation and the opening to the public knowledge of networks that today are obscure to them. A motivation to foster such kind of projects can be found in the Europe 2020 flagship initiative Innovation Union (European Commission, 2010), in the areas targeted by the European Innovation Partnerships (EIPs, see Section 10.3.2.1 below). These partnerships might create new opportunities for doing sustainability research and social innovation, bringing scientists, policy-makers, citizens and civil society organisations, and business stakeholders together in shared processes. Linking participatory processes on complex sustainability challenges to European policy goals can also contribute to enhancing the participants' perception of being "European citizens" – an important and positive side-effect for building European citizenship and social capital – and to bridging the gap between citizens and decision-makers. However, in order to attain a significant impact at EU level, a fundamental requirement would be to institutionalise and systematise the application of the AWARE method. As mentioned above, this would be especially useful for establishing trans-boundary activities and connections between policy-makers, scientists, stakeholders and citizens, for instance at the level of international river basin organisations in the water sector.

The following sections will explore in more detail these two options for achieving a snowball effect.

10.3.1 The AWARE solution for deepening citizens' participation in the water sector

The AWARE pilot experience has been translated into a more general concept of "AWARE solution for participatory assessment of the Water Framework Directive implementation" presented in Figure 10.2.

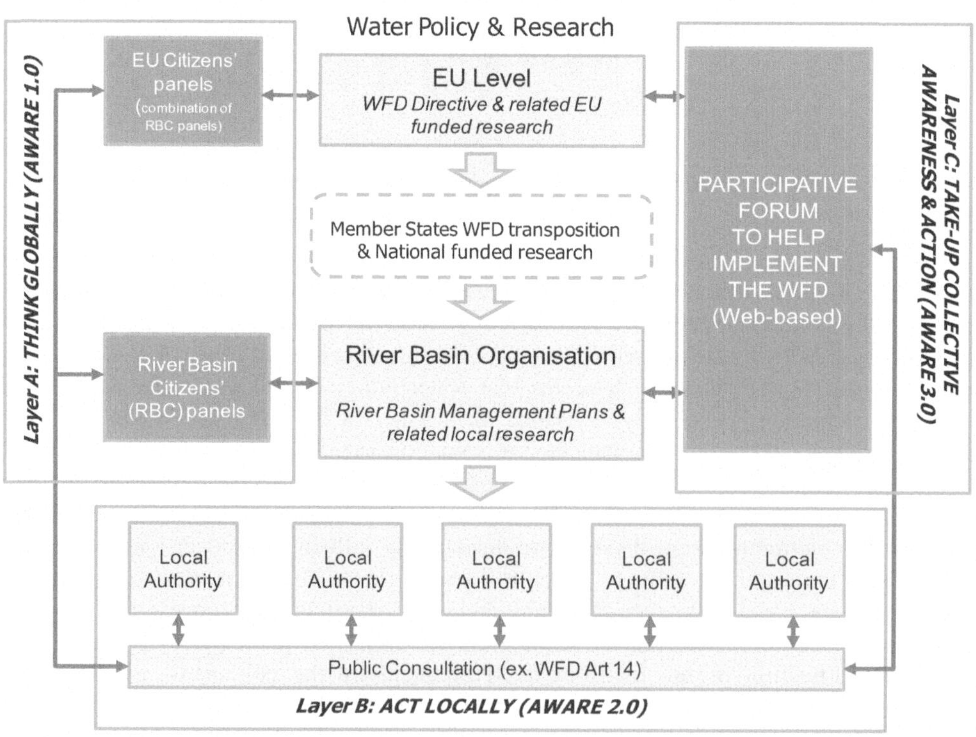

Figure 10.2 The AWARE solution for participatory assessment of the Water Framework Directive implementation.

The AWARE solution considers the EU water policy and research as a whole: from the formulation of the WFD at EU level, through the transposition of the Directive in the Member States (dotted box in Figure 10.2), down to the level of river basin authorities in charge of the River Basin Management Plans, and finally down to the bottom level of local authorities (municipalities) that usually have the responsibility to implement local plans and actions in their territories. The target of all these institutional levels in the framework of the WFD is to achieve a good ecological status of river and coastal waters by the year 2015.

Nowadays, public participation is scarce and mostly organised in the form of public consultations open to local population and stakeholder representatives, and aimed to assess the implementation of local water management plans. There is no formal procedure that would help the citizens to address the water management process earlier, at the stage of formulation of the river basin management plans or, even more ambitiously, at the level of EU policies formulation and start of implementation.

The geographical and institutional distance of the average European citizen from the "Brussels policy centre" has created a large gap, which could be filled in with new participatory assessment procedures and systemic approaches supported by a strong top-down political will at EU level. The AWARE solution, if backed by a consistent political support, may represent such a systemic approach. It may be used to formalise the citizens' participation in the assessment of EU water policy (as well as other policy and research fields addressing grand societal and environmental challenges) by recognizing that there are three distinct layers of participation to be connected:

- **Layer A – Think Globally (AWARE 1.0)**: This is the layer already tested in the AWARE project, that is the participation of panels of randomly selected citizens from different European coastal areas/river basins in a discussion process (workshops and conferences) organised at EU and local levels. The process may engage scientists, lay citizens, stakeholders and policy-makers in a common assessment of (i) water management goals (namely achieving a good ecological status of coastal waters), (ii) the related research evidence, and (iii) the formulation of policies at EU and local level. Therefore, participation in this layer is focused on thinking about (i) global or universal issues – for example deterioration of coastal waters in Europe – and how they are handled by the current policy framework at EU and regional level, (ii) what evidence is produced by research, and (iii) how the brokerage of knowledge between scientists, citizens and policy-makers may help to improve consistently the planning and management of water resources and ecosystems at local level. Layer A envisages the engagement of small groups of randomly selected citizens on sustainability topics of interest, to participate and produce their assessment at an early stage of policy *formulation* (i.e. at a stage where EU regulations and directives are in gestation and an ex-ante impact assessment is needed); or at an early stage of policy *implementation* (i.e. when regulations/directives at EU level are issued and need to be transposed at national level, requiring bottom-up citizens' empowerment approaches to enable an effective implementation in the local contexts). The label *AWARE 1.0* attached to this layer refers to the level of Internet support and interactivity required with the citizens of the panels. Internet is used here in the most traditional manner, to disseminate information about the project, to collect citizens' candidatures, to support the on-going organisation of their participation to the meetings (workshops and conferences) by means of e-mail communications, and to search and collect relevant information from the web as appropriate to support the citizens' work with additional information.
- **Layer B – Act locally (AWARE 2.0)**: This is the layer where more conventional public consultation processes are usually set up, with the aim of building consensus on policies and programs already framed in the context of the River Basin Management Plans. Participation is typically limited in this layer to the ex-post assessment of downstream project details, and related to decisions within the reach and boundaries of single local authorities. This is, however, the level at which citizens' involvement and participation may reach a maximum impact. However, it will only be effective if Layers A and C are applied to enhance the level of knowledge and commitment, not only by lay citizens but also by politicians and stakeholders at the different levels – European, national or regional – where decisions which eventually produce local impacts are taken. The label *AWARE 2.0* attached to this layer refers to the enhanced level of interaction and cooperation of the citizens that may be enabled through the Internet, for instance to support the monitoring of program implementation or provide information useful for policy control and enforcement at local level (for instance, there are several examples of Web 2.0 applications that allow the citizens to quickly and cheaply inform local administrations about the status of the urban environment, roads, etc, by sending self-produced pictures and/or videos).
- **Layer C – Take-up Collective Awareness and Action (AWARE 3.0)**: The third layer is surely the most innovative, and currently still speculative[1], as it aims to enable collective awareness and action, building up online participative forums open to lay citizens, civil society representatives and stakeholders where: i) clear information on the implementation of the Water Framework Directive is provided; ii) participants may open discussion areas and organise discussion events (webinars), as well as develop wiki-pages, to submit recommendations and proposals with the aim to influence the policy debate and law making at EU or river basin level (i.e. beyond the local level). The WFD information could be related to periodic (e.g. annual) impact assessment exercises monitoring the common implementation across Europe. All interested users should be able to provide their comments, discuss policy implementation aspects, publicly control and monitor the progress of policy implementation, and so on. The focus of participation here is on raising and keeping alive the awareness of the public about the implementation of water policy across Europe, and letting the citizens have more permanently the opportunity to

[1]At the best of our knowledge there are no examples of implementation

control its progress and influence policy-making on matters – as the deterioration of coastal water quality – that are relevant for the quality of life. The label *AWARE 3.0* attached to this layer refers to the most advanced form of interaction and cooperation with the citizens and stakeholders that these open platforms should enable, allowing in particular the creation of shared and transparent online groups to lobby or take actions to solve sustainability problems and challenges directly. Although the AWARE evaluation (see Chapter 9) showed that a Facebook page created and shared with the citizen panels received low inputs and resonance, this third layer provides much more opportunity to influence the debate, and to interact not just with other citizens, but across various communities dealing with environmental issues.

Implementing this AWARE solution would be the main avenue to producing an important impact. To achieve this, we would need water managers and policy-makers – at international, EU and river basin level – to appreciate and mainstream the AWARE pilot project results in their water policy environments. We would need also to connect the innovative citizens' conference approach (layer A) tested in the AWARE project with the other two layers, that is local public consultation processes ex Art. 14 of the Water Framework Directive (layer B) and a participative WFD Forum (layer C).

To trigger the above potential impact, the AWARE solution has been disseminated in several scientific and policy conferences during the last project period and it was included as one of the innovative solutions presented at the 6th World Water Forum – Time for Solutions in Marseille in March 2012 (under the topic "Science and Water Policy Interface: When Science and Innovation Meet Water Policies"). Among the main outcomes of the Forum was the intention, supported by several participants, to implement an action plan for improving the science-policy interface, especially in the context of the forthcoming EU Research Programme "Horizon 2020". Amongst the features recommended to frame this action plan, there is the need (Martini, 2012):

- To build multi-stakeholders processes (platforms including research and policy, local authorities, business including SMEs, NGOs, citizens, etc.) at all relevant scales;
- To develop an online platform where citizens can engage in the process to give them the opportunity to influence policies;
- To recognize the importance of knowledge brokers to reconcile the two sides of policy and research;
- To develop financing and incentive mechanisms for research dissemination and uptake, finding in particular funding opportunities and experts specialized in facilitating the transfer of knowledge;
- To ensure knowledge transfer across different geographic locations within consistent regions.

10.3.2 Expanding the AWARE solution beyond the water sector

The ultimate success of a citizen participation process is when it can be shown that the proposals advanced by citizens are translated – even if only in part – into policy. Indeed, one of the first things citizens invited to participate in consultation processes want to know is how the outputs of their discussions will be used and by whom. However, this type of follow-through is often the greatest weakness of citizen participation processes and the reason why such processes are often viewed with mistrust. It is therefore important not only to ensure that deliberative citizens' panels produce outputs but also to document how these outputs are disseminated and, subsequently, used by policy-makers (Move Together, 2009).

The best way that the AWARE successful pilot tests may suggest to better connect science, policy and citizens is to institutionalize the citizens conference processes at EU and local level, by embedding the citizens' jury methodology in a framework of citizens' participation rules adopted at different institutional levels, in relation to sustainability topics which are relevant for applying a structured citizens participation process. For instance, at the local level, there are a number of permanent participatory forums set out in the context of Agenda 21 processes, which are mostly open to and used by stakeholders and civil society organisations, but poorly known or used by lay citizens. These forums usually connect the local administrations with organisations from the local economy and society. They could include also – as a new stream of participation – citizens conferences convened on topics of interest, for example – beyond the water sector – the formulation, implementation and monitoring of Sustainable Energy Action Plans in the context of the EU Covenant of Mayors (www.eumayors.eu).

10.3.2.1 Participatory assessment of EU research to tackle grand societal challenges

The AWARE experience and approach could be therefore expanded in the context of EU funded research, aiming to build up and maintain an effective Science-Citizens-Policy Interface in the whole European Research Area. This could be done by creating a specific "citizens' science" program to fund "European citizens-science projects", engaging together scientists, decision-makers and permanently renewed (through random selection) panels of citizens. Such European citizens-science projects will bring researchers *and* citizens from across Europe together in collaborative research experiences, which today are open only to researchers; such participation would contribute to enhancing the participants' feeling of being truly "European citizens".

One option to scale-up the experience of citizens' workshops and conferences may be through the creation of "*public science-citizens-policy makers interface spaces*" that citizens can access through calls for application to random selections open on the Internet – as was done with the AWARE citizens' recruitment process. Although in these public spaces small groups of people (the selected panels) would be engaged in European citizens-science projects as a voluntary activity, time compensation fees may be distributed to the participants, as was done in AWARE.

While engaged in the face-to-face knowledge brokerage process, the citizens in the selected panels do not rely on the Internet – even cell phones usage is discouraged during the workshops and conference sessions – but are rather involved in an intense exchange of information and opinions with experts, stakeholders and policy-makers. At the end of the process, the citizens' panels co-produce an assessment document ("Citizen Declaration") that can also be disseminated on the Internet, to frame wiki-like User-Generated Content applications (OECD, 2007) and trigger a continuous dynamic in the blogosphere. This approach may evolve with the establishment of "collective awareness platforms" on topics of interest for scientists, citizens and policy-makers, enlarging the arena of participants beyond the panels of randomly selected citizens.

This transition from the "old" to the "new way" of connecting scientists, citizens and policy and market developments is envisioned in Figure 10.3. In the conventional way, EU research funding contributes to the development of Science & Technology (S&T) in Europe. Broadly, total Research & Innovation funding in Europe reaches about 200 billion euro per year, and EU funding makes up about 15 billion euro of this total (European Commission, 2008). In turn, S&T supports policy-making by helping decision-makers to deliver better government decisions. S&T may also eventually improve or produce new commercial technologies, products and services sold on the market. Citizens in this traditional picture are mainly seen as "customers" of private companies or "end users" of public services; they do not have enough opportunity to *interact and contribute early in the process* and *influence* the way public government and the private sector tackle societal challenges, in which citizens have a significant stake and interest.

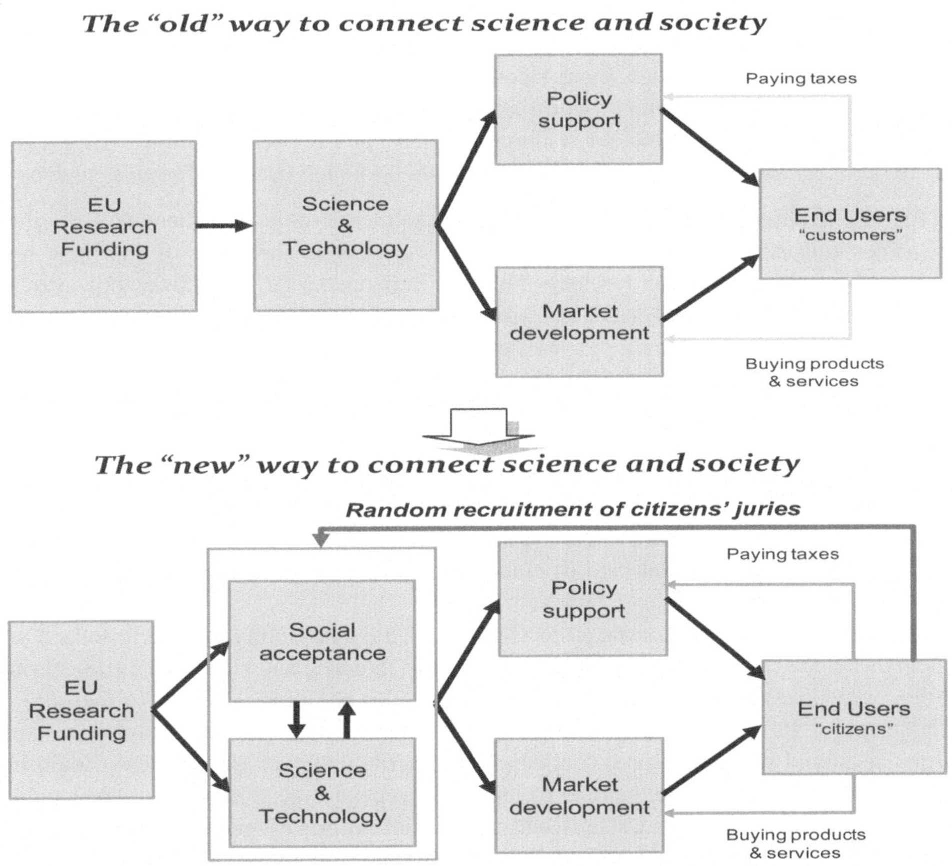

Figure 10.3 A new way to connect science and society by means of participatory assessment of societal acceptance.

The "new way" of connectivity adds an opportunity for the citizens: as part of randomly recruited citizen juries, they can make a critical review and assess the social acceptance of the research funded by the European Union, in particular research addressing grand societal challenges. This could be framed as a formal citizens' commitment – a kind of "European civic service" – which in addition to increasing their sense of common European citizenship also broadens the set of tools, knowledge, and ideas with which policy-makers and researchers are working. Europe-wide societal challenges that can be focused on, for instance, the ageing population, the effects of climate change, and the growing scarcity of natural resources (energy, water, land or raw materials). The Europe 2020 flagship initiative Innovation Union (European Commission, 2010) called for linking future EU funding programs more closely to these topics, which will clearly affect future European citizens' health, well-being and quality of life.

The main instruments to connect EU funded Research & Innovation – the business world and civil society – are the European Innovation Partnerships (EIPs) launched as part of the Innovation Union flagship initiative. EIPs are going to be launched only in areas – and consist only of activities – in which government intervention is clearly justified and where combining the EU, national and regional efforts in Research & Innovation and demand-side policies will achieve the target quicker and more efficiently. Examples of partnerships under consideration first address areas of great concern for European citizens:

- *Active and healthy ageing*: by 2020, to enable European citizens to live longer in good health by increasing the average number of healthy life years by two, and, in achieving this target, to improve the sustainability and efficiency of our social and healthcare systems.
- *Smart and liveable cities*: by 2020 the aim is to support a number of pioneering European cities in reducing their carbon emissions by more than 20%, increasing the share of renewable energy by 20% and increasing end-use energy efficiency by 20%. The partnership aims to demonstrate that sustainable development of local economies, with a smarter use of scarce energy and natural resources while continuing to improve the citizens' quality of life, is possible.
- *Water-efficient Europe*: to promote actions that can speed-up innovation in the water sector and remove barriers to innovation, ensuring higher quality of water and a more efficient use.
- *Smart mobility for Europe's citizens and businesses*: to equip Europe with seamless door-to-door travel and effective logistics by promoting the broad and coordinated development of Intelligent Transport Systems, and building upon available results from research and development to take them further to innovation and concrete operational deployment. Together with needed behavioural changes, this will contribute to achieving better mobility and transport in Europe, reducing congestion, energy consumption and transport-related greenhouse gases emissions.
- *Agriculture productivity and sustainability*: world food demand will increase massively over the next two decades. The aim of this partnership is to promote a resource-efficient, productive and low-emission agricultural sector – which works in harmony with the essential natural resources on which farming depends, such as oil and water.

The EIPs are thus the new avenues indicated for future research to tackle societal challenges. Bringing researchers and stakeholders from across Europe together in collaborative networks is at the heart of this new approach and will continue to be vital in sustaining a European research fabric. Experience so far has shown, however, the limitations in achieving the necessary flexibility, creativity and cross-disciplinary research needed. Moreover, there is an increasing need to get Europe and the European research – at least to the extent that this is going to change the citizens' life in a foreseeable future – into the minds and lives of ordinary citizens.

How could this be achieved? One way would be to introduce in the next EU Common Research Framework "Horizon 2020" *a target of 10% of Research & Innovation funding* allocated for European citizens-science activities, mainly in the context of the new European Innovation Partnerships. This would mean to allocate an order of magnitude of 1,5 *billion euros* (about 3 euros per citizen per year) for European citizens-science activities in the new Financial Framework 2014–2020. Those European citizens-science projects should include as eligible criteria *transnational panels of citizens*, whose members will have to be randomly selected from at least 3 different countries of Europe, applying here the same rule holding today for forming European research consortia.

Citizens involved in European science-citizens projects will have the opportunity to meet in several workshops around Europe, and work together with scientific experts, other involved stakeholders, and with other citizens from other countries. While doing this, they will have the opportunity to exercise English as common language (*a learning by doing* process improving their language skills), and to appreciate the different national cultures, exploring commonalities and differences, in particular around specific societal challenges as described above. This process should be open to people from all ages, social status and walks of life, helping to build a European "social capital" of cross-country relationships and friendship networks. In addition, as the citizens will acquire in the process adequate scientific information on the sustainability topics of concern, they will gain new skills (e.g. team working), and a better understanding and awareness of the complexity of societal challenges.

Other supporting actions connected to the proposed "EU citizens-science program" may include life-long learning and training activities aiming to enhance the citizens' common language and other civic skills, for instance by means of "*European citizens' summer schools*", as well as the skills of facilitators of participatory projects. Citizens' summer schools can be organised by networks of universities and training institutes around Europe, with the support of other regional or local institutions. Applicant European citizens would be trained on EU sustainability policy frameworks and related science issues, and citizens' participation methods and tools. They will have the opportunity to share an international "classroom" environment by working together with experts, for instance to assess the societal acceptance of science topics related to their everyday life concerns. During the AWARE project it was noted that as several participant citizens gained expert knowledge and information, they planned to take this back to their colleagues, families and communities, thus snow-balling the exchange they had begun during AWARE.

10.4 PRACTICAL LESSONS FOR FUTURE PARTICIPATORY PROCESSES

The main conclusion from the AWARE pilot experience is that knowledge brokerage processes based on proven methodologies can be used systematically to improve the science-citizen-policy interface in Europe, supporting decision-making on sustainable water management and other sustainability issues. A set of practical lessons, based on the AWARE experience, was finally delivered as a guidance for managers and decision-makers that are applying – or are willing to apply – participatory processes (see also AWARE, 2011b). The lessons are related to engaging in a knowledge brokerage process, respectively, citizens, scientists, stakeholders and policy-makers, and how to organise effectively the knowledge brokerage and ensure that outcomes are delivered. They are presented in the following sections, as tips for future applications of the AWARE method.

10.4.1 Tips about engaging citizens

- Instead of traditional open hearings a better response and feedback may be gathered through a random selection of individuals that form a citizen panel – these should then be part of the entire policy consultation process. This requires a careful selection procedure using a call for citizens appropriately disseminated in print and online to the target audience; the collection and evaluation of citizens' application forms; and the selection of panel members and deputies from the pool of candidatures received with the support of software ensuring a fair opportunity to be selected and a balanced composition of the panel (e.g. in terms of age, sex, activity, attitude and motivation towards the topic).
- The commitment of the citizens selected as panellists need to be ensured at the very beginning of the process, by signing a letter of commitment where the terms and conditions for their participation are established (usually to attend workshops and conferences at fixed dates) and a nominal fee to compensate for their time (about 6 to 10 days over one year) is agreed to be paid at the end of the process. Any travel and accommodation costs needed to attend transnational workshops must be covered from the project budget.
- Citizen panels should aim to be representative of the socio-economic structure of the case study they represent, however considerations about language proficiency and a basic level of interest in the sustainability topic addressed are a priority. The ability to speak a common language is crucial when working with transboundary and cross-European citizen panels: the presence of language interpreters would greatly reduce the effectiveness of interactions between participants and would significantly increase costs. Substantial time however, is still needed to clarify for those involved the terminology of relevant environmental laws and directives, as well as the meaning of relevant scientific terms.

10.4.2 Tips about engaging stakeholders and policy-makers

- Engaging stakeholders from across the low-high interest and low-high influence continuums is crucial in order to achieve a balanced exchange of knowledge, views, and information.
- The participatory process gains credibility by tasking scientists and trusted regional NGOs with the stakeholder engagement.
- The participation of a permanent "Policy and Science Advisory Group", as in AWARE, can provide significant feedback and positive inputs both during the knowledge brokerage events and during the evaluation. Members of this group should be key actors in the study areas; they should also have a relatively high interest in the process, and come from different backgrounds.
- Industry representatives are a key actor – when they are missing from the discussion a wide array of needed knowledge is lost, which has repercussions on the process and outcomes. Reaching this target audience in future projects may involve conducting bilateral consultations with industry representatives around concrete outcomes and recommendations.
- A suggested alternative to enable a more productive public-policy interaction is to engender a *continuous informal science-citizens-policy-makers interface* process of consultation on key sustainability issues. Such an awareness raising process could help bridging the gap between the citizens' locally specific and experiential knowledge, perspective and understanding of the topic, that of elected representatives, and the more technical perspectives of the water managers and the experts involved in the water policy formulation process.

10.4.3 Tips about engaging scientists

- Citizens' input can help scientists to focus on a more comprehensive view of the problem at stake, avoiding the pitfalls of compartmentalisation.
- Including the opinions of stakeholders and citizens enriches scientific models and scenarios and helps develop more robust results. Systematic approaches should thus be developed to promote this type of interaction.
- Citizen-scientist interactions benefit from a regular consultation process across time, during which knowledge and information can be exchanged, trust can be built, and a "common language" based on understanding of complex challenges and mutual awareness can be developed.
- Complementing participatory workshop interactions with public conferences helps maintaining the actors' motivation and interest in the process, and provides an ideal public forum for the presentation of the achieved results as well as the opportunity to build up a consultation around the citizens' recommendations.

10.4.4 Tips about organising the knowledge brokerage process

- Good moderation by a team (moderator plus assistant) of communication or social experts that is trusted and involved throughout the participatory process is essential.
- Citizens' panels need time to consult not only with experts during workshops and conferences, but also among themselves, if they are to produce concise, concrete, and cohesive statements on a given sustainability issue.
- Facts and figures presented by experts should be easily understood and available; sufficient time should be allotted for digesting the facts presented, and clarification questions should be encouraged.
- Expert knowledge should be shared and embedded continuously into the knowledge brokerage process. In order to better convey complete scientific information to the participants, experts could be involved throughout all events, serving as a "knowledge repertoire".

10.4.5 Tips about delivering outcomes

- In addressing complex sustainability issues, the outcomes of a knowledge brokerage process are closely affected by the extent to which the whole spectrum of actors is involved.
- In a knowledge brokerage process engaging citizens' panels, it is necessary to address expectations regarding the outcomes of the process, and the concrete possibilities of implementing certain options. The willingness of individual policy-makers to communicate openly and take up insights from a body of lay citizens is a key prerequisite to achieving concrete impacts on policy processes and decisions. However, an important barrier is the mismatch that often occurs between the responsibility of the local policy-makers (e.g. municipal councillors) to take care of concrete actions at the local level and the public budget that they directly govern, which is often smaller than the budget required to cover the actions focused and deliberated in the participatory processes. Higher levels of government (regional or national bodies) should often intervene, whenever the actions require larger budget allocation, so it is important to gather their full commitment early in the participatory process.
- All the participatory events should be planned to maximise their communication effectiveness, including clear definition of roles and objectives. Facilitation should guarantee inclusion and proper balance of all the participants, time for discussions, understandable information, appropriate and respectful ways of interaction. Giving ample time for the participants to consult in the co-creation process is crucial for a cohesive outcome (Declaration) that has the full support and ownership of the citizen panel.
- Achieving a common basic knowledge of the issue at stake and using a commonly-understood language are key aspects to bring all actors at the same level and thus to develop synergies. Informal types of knowledge may be accepted by some participants (e.g. lay citizens) while more formal knowledge is required by others (e.g. implementers). One effective solution proven in AWARE was to develop a multi-language glossary of technical terms (in this case related to eutrophication phenomena) to help the citizens' panels in understanding and comparing different terms.

REFERENCES

AWARE (2011a). Deliverable 2.8–Report of the EU Citizens and Stakeholder Conference on Linking Research to Policy in the Water Sector, EC FP7 Project, Brussels, available online at http://www.aware-eu.net/ (accessed 10 April 2012).

AWARE (2011b). Deliverable 4.3 – Bridging the Science Policy Gap: Best Practices in Citizens' Participation, EC FP7 Project, Brussels, available online at http://www.aware-eu.net/ (accessed 10 April 2012).

AWARE (2011c). Deliverable 4.8 – EU level interviews results, EC FP7 Project, Brussels, available online at http://www.aware-eu.net/ (accessed 10 April 2012).

European Commission (2008). A More Research-Intensive and Integrated European Research Area. Science, Technology and Competitiveness key figures report 2008/2009, Directorate-General for Research, European Research Area, EUR 23608 EN.

European Commission (2010). Communication from the Commission to the European Parliament, the Council, the European Economic and Social Committee and the Committee of the Regions: Europe 2020 Flagship Initiative Innovation Union. COM(2010) 546 final, Brussels, 6.10.2010.

Martini F. (2012). Final Target & Solutions Group report Complement from the joint CS3.1/EU10 session "Science and Water Policy Interface: When Science and Innovation Meet Water Policy" Main Outcomes, 6th World Water Forum, Marseille.

Move Together (2009). Guidelines for Local Authorities for the Organization of Citizen Participation, EC FP7 Project, Brussels, available at www.move-together.net (accessed 10 April 2012).

OECD (2007). Participative Web and User-Created Content: Web 2.0, Wikis and Social Networking. OECD, Paris, p.124.

Shirky C. (2008). Here Comes Everybody: How Change Happens When People Come Together. Penguin Books, London.

Appendix

AWARE: Connecting people for better water management

EUROPEAN CITIZENS' DECLARATION

For a sustainable management of coastal water ecosystems in Europe

Arianna BERNARDI, *Italy - Ferrara*

Arnis BERZINS, *Latvia - Riga*

Fabio CAMISOTTI, *Italy - Corbola*

Aija CAUNE, *Latvia - Riga*

Nicolas COPIN, *France - Lille*

Veronique DE RIDDER, *Belgium - Grimbergen*

Mikhaëlla FIEL, *France - Paris*

Beatrice GOFFARD, *Belgium - Andrimont*

Gilda GORI, *Italy - Goro*

Giovanna GORI, *Italy - Goro*

Emmanuel HANKENNE, *Belgium - Brussels*

Aivars HINCS, *Latvia - Ogre*

Agnese JEKABSONE, Latvia - Sloka

Karin KOPS, *Estonia - Tartu*

Alain LEDAGUENEL, *France - Dunkerque*

Ann LEFEBVRE, *Belgium - Namur*

Arianna LONATI, *Italy - Bosco Mesola*

Claudio MANFREDINI, *Italy - Ferrara*

Jean-Francois MASSELOT, *France - Lille*

Anda MIZE, *Latvia - Riga*

Isabelle MORE, *Belgium - Brussels*

Aleksandrs PETJUKEVICH, *Latvia - Daugavpils*

Gina PIEGAVA, *Latvia - Riga*

Roberto POLETTI, *Italy - Ferrara*

Ieva REINHOLDE, *Latvia - Riga*

Kaur SARV, *Estonia -Tallinn*

Angelo SCARPA, *Italy - Goro*

Maria Cristina VERATELLI, *Italy - Porto Garibaldi*

Diego VIVIANI, *Italy - Goro*

Benjamin WOUTERS, *Belgium - Brussels*

What is the Aware Project?

The AWARE project (www.aware-eu.net) is funded by the Seventh Framework programme of the European Commission. This original European initiative jointly engages scientists, policy makers and citizens in creating and analysing scenarios for a sustainable management of coastal water ecosystems in three European areas:

- The Gulf of Riga (Estonia and Latvia),
- The Southern North Sea and Seine, Somme and Scheldt river basins (France and Belgium) and
- The Po River Delta (Goro lagoon, Italy).

During the course of the project, citizens, scientists, policy makers and other stakeholders concerned by human-caused deterioration of coastal ecosystems actively participate in workshops held at both local and European levels. AWARE's approach is collaborative and solution oriented, focusing on the dialogue between the key actors to improve the management of the situation.

The AWARE project aims at delivering recommendations and scenarios to European and local policy makers, and ultimately making all stakeholders aware of the urgency for a sustainable management of European coastal water ecosystems.

What is Eutrophication?

Eutrophication is a process driven by enrichment of water by nutrients, especially compounds of nitrogen and/or phosphorus, leading to: increased biomass of algae, changes in the balance of organisms, and water quality degradation. Eutrophication is a natural process that has intensified due to its extreme sensitivity to human activities and their effluents (excess nutrients) discharged from *point sources* of pollution (e.g. sewage and waste water treatment plant) and *diffuse sources* of pollution (e.g. leaching and run-off from agricultural soils). These problems have not only environmental but also serious social and economic consequences on the impacted areas.

THE THREE DISTINCT AREAS OF STUDY

The Southern North Sea and Seine, Somme and Scheldt river basins area

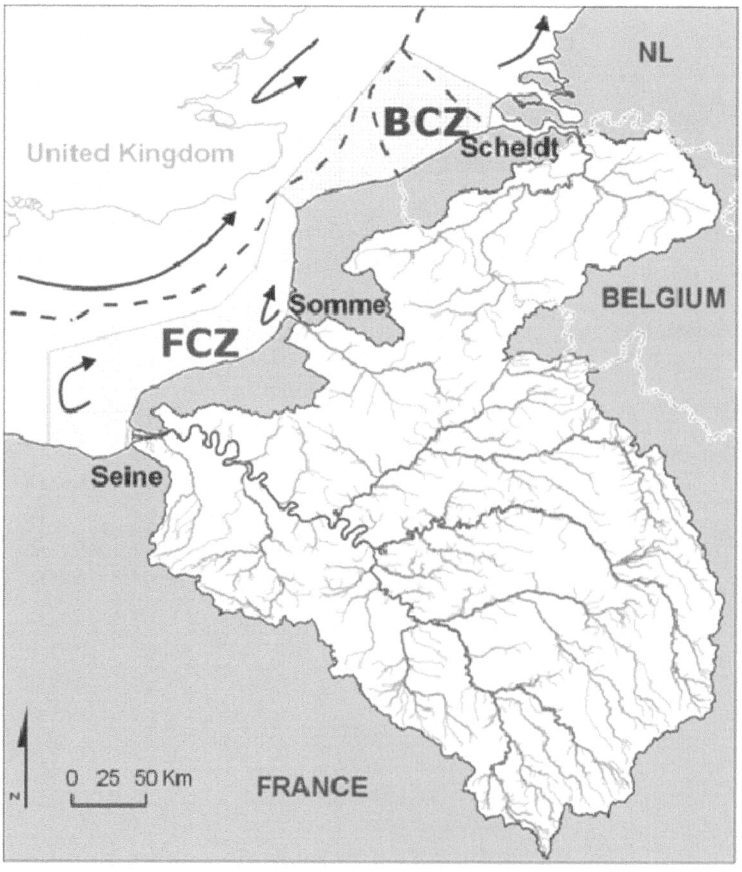

The Po River Delta and the Goro lagoon

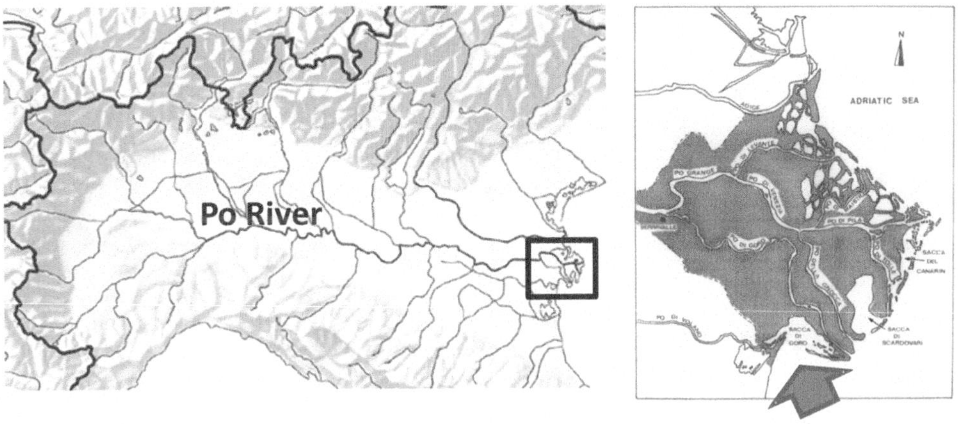

The Gulf of Riga

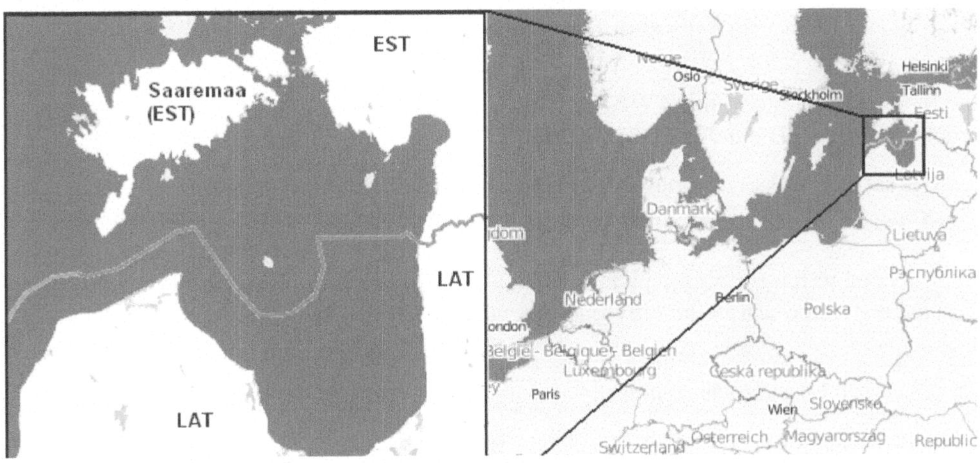

INTRODUCTION

Who we are

We are a group of 30 citizens involved in the AWARE European project. We had volunteered and were selected randomly to participate to a project that aims at improving coastal water management by leaning on a better connectivity between research, policy making and the civil society. We come from three European areas:

- The Sacca di Goro lagoon (Italy – 10 citizens);
- The Gulf of Riga (Estonia and Latvia – 10 citizens);
- The Southern North Sea and the Seine, Somme and Scheldt river basins (France and Belgium – 10 citizens).

We all have different life styles and situations but we have a common interest, that is, we share the same will to understand the situation and play an active part in improving it. We are here today as citizens; we are not directly involved in water or land management decisions and we are not specialists. However, we now understand much more on the issue thanks to the discussions among us, and with scientists and other stakeholders at local and European levels.

How we worked

It all started for us in Paris in April 2010. We initiated debates on our perceptions of coastal water quality in each of the three European sites. The topic of particular interest was eutrophication, its causes and consequences, and the European legal framework that drives coastal ecosystem management.

At the end of 2010/early 2011, we organised local workshops to study more specifically eutrophication and other water quality problems in our area. Eventually we were seeking to elaborate local recommendations for a better management of our coastal waters.

In April 2011, all 30 of us gathered in Paris to share our conclusions from our local level workshops, and put them in a European perspective.

How we understood our "mission" in the aware project

Our objective was to provide our points of view and recommendations after broad debates on how to improve the management of coastal water ecosystems through better connectivity. That involves, among all the concerned actors, us as citizens.

The project focuses especially on the phenomenon of eutrophication, as a marker of water quality. Although it is a valid marker, our first impression is that *eutrophication is only one entrance door* to the broader issue of ecosystem health, including other sources of pollution, like Persistent Organic Pollutants (POP), potential radioactivity, biological pollutions (i.e. germs and microbes), heavy metals, synthetic molecules ("biocides", hormones…), oil and derivatives, coming from BOTH the land AND the sea.

What were our main concerns as citizens regarding the quality of water in Europe?

- Even if we did not necessarily understand where the problems come from, what is really happening, or what we could do about it, *we are concerned about the environment we live in* (the one we are borrowing from our children). The impact that bad water quality could have on babies and future generation is definitely in our mind.

Sometimes we observe foam on the beach or can't see our feet through the water while walking on the beach, and we didn't understand why.

- Besides, we have realized along the AWARE process that *the issue of water quality cannot be restricted to the eutrophication problem. It is much more complex.*

Multiple sources are involved: industries, households, tourism, agriculture, other countries upstream - sometimes beyond the border of the European Union (EU), from the sea because of the marine activities (transport, fishing…), etc.

⇒ We think it is needed to have a bigger picture to be able *to apprehend the issue in its entirety*; we feel a holistic (or multidimensional) approach is the only one that can help understand the water quality issue, and set the way to solve it.

- We are also concerned by the *issue of time*:

We are worried by the observations showing the effects of several years of (poor) water management onto wildlife deterioration (on seabirds, zooplankton and phytoplankton, plants affected…). Protected species in the Po Delta are affected by eutrophication, for example. We are also worried about algae blooms being toxic or about polluted water, causing sickness to sea shore users (children being especially vulnerable). We fear it might have long-term impact with consequences on human health that we cannot even completely foresee today.

⇒ We know that we are paying for several decades of human practice and that the ecosystem has a strong capacity of inertia: change will come slowly.

That's why it is not time for talks anymore, it is time for actions.

- We observed in the three regions that the environmental and economical issues are sometimes in contradiction. *A balance between socioeconomic aspects and the environment is needed*, in building scenarios for an improvement of the situation.

Indeed, in Sacca di Goro, as local people rely heavily on the state of their lagoon to produce their major source of revenue (clams' farming), the AWARE citizens of this area call for solutions which will combine the regulation of impacts on the environment and reduction of eutrophication as long as fishers' jobs are not affected.

In the North Sea case, the AWARE citizens are very attentive to the importance of balancing priorities to support the development of the right infrastructure to reach the objectives of the Water Framework Directive and to maintain and develop economic activities in a sustainable way.

In the Gulf of Riga, the local AWARE citizens call for a balance between ambitious ecological goals and social needs and acceptable costs in the search for solutions and improvement scenarios. They were worried of the cost effectiveness of the solutions.

⇒ *No option is the best per se. It has to be evaluated against environmental and socioeconomic criteria.* But we trust that acting now is always cheaper than standing still. These criteria have to be taken into account at the coast but also more upstream, where pollution is often produced and has an impact on coastal waters.

What did we experience in aware?

In the time we have spent participating in the AWARE project we have found out how important it is to connect scientific knowledge and water management practices to achieve a good ecological status of the coastal waters. Although co-operating with citizens from the other two case studies, we have found out that while in all three case study locations we are encountering similar problems, and their possible solutions seem similar, they should be treated each with its own unique approach and different criteria.

For example in the Gulf of Riga case we are encountering a too high level of phosphorus which is – as we understood – mainly a "heritage" of bad status and insufficient quantity of waste water treatment plants (WWTP) as well as ex-USSR army presence. At the same time, in North Sea case, the level of phosphorus is acceptable, and waste water treatment system is fine but there is a too high level of nitrogen, which call for solutions on diffuse sources of pollution, like agriculture practices.

Eutrophication was a concept unknown by most of us citizens, prior to our participation to this project. By interacting with scientists having detailed knowledge on the topic, and learning from different channels (i.e. talking to people on the ground, farmers, fishermen, managers of waste water treatment plants, decision makers, etc.), we as citizens have been able to acquire an expertise on the subject of matter. At the same time we realize that many citizens actually have no idea about it and hearing words like eutrophication will not say much to them.

Our rich discussions with scientists have shown us the importance of relying on different sources of scientific knowledge. The dialogue between us – lay citizens- and them revealed that they are kind of "sentries" or "watchmen" who can warn on the evolution of a situation, and play a role of keeping the citizens alert, informed and aware, while we can provide them with our opinion on priorities for testing scenarios of solutions to improve the situation. Our perception of citizen can also ensure that any aspect has been taken into account in a global perspective of the situation – as already said before. This way, the information can be shared between citizens, with their knowledge of the local situation, and scientists, with their in depth knowledge of details.

During the process of our co-operation in Aware we have proved that lay citizens from three different countries can work successfully on three different cases with diverse local conditions, support each other, and achieve common goals and mutual informative enrichment.

While listening to visiting guest lecturers we have also encountered situations when different stakeholders are in fact struggling. Each stakeholder is representing and defending his own interest that does not always match with the others'. It is necessary to find a way to balance their interests because changes in their current actions cannot be made without paying a part of the cost … and no changes will be way more expensive.

For example when we were talking about possible water pollution level as a result of different agricultural actions, its control and ways to reduce it, an invited representative of Copa-Cogeca was claiming that this is not possible without significant increase in end-user prices of agricultural products, while at the same time members of AWARE personally have been visiting individual farmers who have in fact already achieved a conversion to sustainable farming practices reducing impact on water pollution.

While we were investigating this and related issues we have found that as more information about subject we are getting the more additional aspects (which also must be considered) appears. For example if we look at the situation in the Po river basin[1]: the best composition of water in the Sacca di Goro for the breeding of clams is not necessarily suitable with the ecological quality of waters required by the EU Water Framework Directive. Moreover, we discovered that the level of pollution can be seasonal: in summer, when many agricultural activities require high level of water drainage from the river, the level of pollution rises to a much higher amount because overall quantity of water reduces seriously; whereas, the same river can be considered with a normal level of pollution in winter (also due to extra water inflow into river from snow).

Why are we here?

- ***Responsibility***
  We see there are environmental problems around us, and we cannot always understand them. For instance, there are regularly huge amounts of foam on the southern North Sea coast around spring, a sign that there are problems, and we didn't understand the reasons. This is one of many illustrations of the issues each of us can see on a regular basis. *We recognize our share of responsibility for the impact on the environment and therefore we are willing to act.*

You hear good intentions all the time; but good intentions will not create the change required to improve the quality of coastal waters; real actions will.

- ***Carrying our message across***
  We want to be heard as citizens. Scientists and authorities too often make the decisions without considering citizens' opinions but we have our own expertise, and we believe we have the rights to express our concerns. We want to be

[1]Po is the biggest river in Italy, which supplies water to the Sacca di Goro lagoon. In its basin are living about 16 million people.

considered on the same level as scientists and other stakeholders. For that we need to be informed, we need to understand what the options are and how best to contribute. We are ready to be part of a change, but it is hard for individuals to be heard when you do not belong to any representative organization. When you do, you have to carry its voice and can lose the opportunity to express your personal opinion.

What are we ready to do as citizens?

Individuals are key to the water ecosystem management, from a wide range of positions: as observers of the situation and direct users of the sea and water, as consumers of the products that can degrade the environment upstream, as citizens part of the decision-making process, alone or collectively, through an association or by taking part to a participatory process such as AWARE, and so on.

- ***As consumer, we are willing to take actions***: that is acting as enlightened consumers, analyzing how what we buy is being produced.
  The changes we'd like to see in agricultural practices will come if the consumer practices evolve. We have learned for example that it takes 7 kg of cereals to produce 1 kg of meat; if not stopping our meat consumption, at least we are keen to reduce it. Reducing animal product consumption (and production) by a factor of two, would allow indeed much less intensive agricultural production hence, lower nitrogen pollution.
- ***We will be critical about the data*** as we now have more knowledge on how to interpret it.
- ***We are willing to share our expertise*** with scientists/stakeholders/policy-makers/other citizens.
  We are willing to convey a message at all levels, which includes sharing the experience we gained on this project, and take home with us a message that will foster a change. For decisions and recommendations that would fit, we will even be implementers at our level (at home, at work, in our society).
- ***We will review and track promises*** by the politicians, check for short term commitments and demand tracking results of long term commitments.
  In the declaration below you will not find only recommendations we address to policy-makers, scientists and stakeholders. As we understood we share a part of the responsibility on the whole process and management of water ecosystems, as citizens, we are also ready and willing to act. This is why we are here, and we are going to share with you our own views on the situation.

I. HOW CAN THE SITUATION CHANGE? HOW CAN WE ACHIEVE A BETTER AND SUSTAINABLE COASTAL WATER MANAGEMENT IN EUROPE?

Solutions exist. We have discussed options for improvement in the framework of the European water legislation, and we have seen management practices implemented on the ground in each case study. We discussed these all with one goal in mind: achieving good ecological status of coastal waters by 2015, called in by the European Water Framework Directive (WFD).

How can the situation improve thanks to policies at the European level?

How can the situation improve concretely on the ground?

1) Eu classification of water quality: About "good status" and the implementation of the WFD

Our findings

After receiving a presentation about what is a good status for water quality[2], it appeared to us that the official scale of measure (from bad to high quality) for water quality is not so clear and not so efficient.

Status	Code	
High		"Reference condition" Acceptable ecological status
Good		
Moderat		Action needed for achieving acceptable ecological status
Poor		
Bad		

[2]The EU Water Framework Directive is the legislative tools for the management of waters. It commits European Member States to achieve good status of all water bodies (inland, transitional and coastal) by 2015. To achieve "good surface water status" both the ecological status and the chemical status of a surface water body need to be at least "good".

Why? We see:

- ***No common evaluation system*** (not the same scale and quotation system for every European Union member states) and *no consistency of data across Europe*.

Indeed, every EU member state can define its own view/analysis of what is a good level of quality of water. This process is considered by us as a risk to level down the scale quotation. How can we trust the data if there is no consistency in Europe? It was a shock for us to discover that there is no common understanding of EU member states on how to monitor the quality of water. In the Riga area for example, there is a river crossing two countries: does it make sense if a country says the quality status is "good" while, on the other side of the border the status is said "moderate"? We think there should be a common understanding of the measurement and each member state should not decide on its own indicators what is considered as good quality level.

This does not mean that we must establish a common system throughout Europe – but a common framework of understanding. A high rate of an element can be negative in a local system and positive in another one, taken into account local specificities.

- ***No functional indicators*** (indicators connected to the real functioning of the system and its complexity, where not only static data are taken into account but also fluxes and the way the system works, the way data are combined).
- ***No socioeconomic local analysis*** (social analysis, economic analysis, cultural analysis, particularity of a place analysis…).
- ***No links with outside topics analysis*** (analysis of the soil, analysis of the agricultural sector, analysis of the industrial impacts…).

After reflection, we consider that the implementation of the WFD has done a huge step towards the improvement of the water quality in Europe but, because there is always a "but", this is the time to enter in a new era by thinking to a new process.

In 2000, according to what we understood, the WFD has been implemented as followed (in chronological order):

- The writing of the Directive;
- The definition of the objectives and indicators (which were structured indicators);
- The establishment of the scale and the national "interpretation" of it and the "liberty" to consider or not indicators;
- The action taken by EU member states or the inaction (consequence of a non-action means a fine for the concerned member state, but we fear that sometimes, paying an annual fine is less expensive than invest money and develop projects.);
- 2015: the water will be at least at the "good" status.

Our recommendations

After many discussions with scientists, stakeholders, citizens, policy and decision-makers, we realized that *there has been little room given to citizens so far in the implementation of the WFD*, and that the scientific work that has been done does not seem complete to us.

Even if we missed some time to go further into our analysis, we have imagined some elements for an update of the current process:

- ***Independent scientists*** (like scientists in national scientific research bodies – FNRS in Belgium, CNRS in France, etc.) analyze different local places and create structured and functional indicators (economic, biologic, chemicals, cultural, etc.).

In our view, to avoid the risk of inconsistency of data throughout Europe, some check should be carried out beforehand by independent scientists to determine whether the measure is decided impartially by a member state.

This independent analysis could take into account some specific local differences between systems (the concentration of an item may be beneficial in one location and detrimental to another). Socioeconomic aspects may be taken into account and even considered as important as the environmental aspects around the issue of the water quality.

- ***Scientists confront their point of view to a panel of lay citizens*** to allow non-measurable issues to appear and try to model those new issues.

  We think we are legitimate to be in the loop from this early stage to share our opinion.
- ***Scientists draw a new scale for measuring the quality and define the status of water.***

  It should be linked with other analyses as water and agricultural analysis, and EU policies like the Common Agricultural Policy (CAP). This scale must be used by every EU member states and interpreted in only one way on a same river basin. Being more open to other analyses, this new measurement of the quality of water would thus offer a more global view of the issue (keeping in mind that the analysis must take into account and identify particularities of certain places).

- ***European institutions and the EU member states provide efficient means to support the independent scientific and global analyses***, which will complement the studies of local scientists and would bring together the different existing studies.
- An updated Water Framework Directive is written.
- The EU supports a stronger *continuing evaluation system*: regular analysis and feedbacks are made to the EU *and citizens* (who can raise new issues appearing), allowing *continuous improvement and strict control* of the implementation of the Directive.
- The achievement of the goal, and the setting of new short-term and long-term objectives.

⇒ In brief, we recommend improving the *consistency* in the measurement of data (instead of interpretation), thanks to independent scientific interventions and broader analyses of a situation. We also recommend a kind of "bottom-up" approach, which would allow the acceptance of the particularity of a local place and thus, the *adaptation* of the scale of measurement of the quality status of water.

⇒ Finally, we think it is necessary to achieve more consistency between different EU policies, like the Water Framework Directive and respectively the Common Agricultural Policy and the Marine Strategy, as it has started. Could DG Environment of the European Commission, as working on a transversal topic, be playing a bigger role in providing information or in making bridges between those different policies?

2) Improvements on the ground: Acting on point and diffuse sources of pollution

Our findings

We understood that eutrophication in the three case study regions resulted from excess nutrients being discharged from point sources (e.g. sewage and waste water treatment plant) and diffuse sources (leaching and run off from agricultural soils). However their relative contribution to pollution loads is not the same in the three regions. We are conscious that actions can be taken to reduce pollution from diffuse sources and point sources, and they should be treated with a territorial approach.

For example, the agriculture in the Gulf of Riga is not as intensive as in the North Sea area studied, and the main problem is the amount of phosphorus coming from point sources which can be eliminated by waste water treatment plants.

In the North Sea case, it seems to us that the main measures to reduce eutrophication should focus on diffuse sources, including agriculture. We wonder whether investments to address point sources pollution aren't coming close to an unreasonable point (both in terms of cost and processing techniques). It is therefore a priority for us to allocate resources to fight against diffuse sources including agriculture.

This is also the case for the Sacca di Goro, as the main source of nitrogen is actually agriculture.

Especially for the Riga case, we expressed the importance of including neighbouring countries outside the European Union with incentives to cooperate in accessing good water quality and monitoring it.

Our recommendations

- ***Enhance works and display information of their positive results to promote organic and other sustainable farming practices.***
  Indeed, as the scientists presented to us, the conversion to another type of agriculture can have a very positive effect on water quality as it reduces the algae blooming and it reduces the need to import soya for example. A French farmer explained to us that after converting 1/3 of his crops into organic farming, he measured improvements in water quality around his fields and that it was cost efficient because he bought less fertilizers and pesticides.

  The conversion to an alternative agriculture should be territory based. For instance, in some territories, organic farming could work while in other ones it would not, depending on the intrinsic attributes of the ground, i.e. near points of abstraction of drinking water for example).
- As citizens we should *adapt our consumption, consume less meat* (to produce 1 kg of meat, the farmer needs 7 kg of cereals) so that when consumption changes intensive farming can be reduced and farming impact on eutrophication lowered.

  Citizens can change their priority and opt for organic production. For example in Goro, we recommended the local authorities to make the inhabitants and tourists aware that they are responsible for the quality of what they eat.

  We also strongly urge for the *need to enhance social ties with farmers and reinforce dialogue* rather than pointing the finger at them for supposedly being responsible of overall environmental impacts. They produce the food we buy and eat. Therefore, it is a shared responsibility.
- In order to improve the water courses, there should be *maintenance of natural wetlands and reed beds* which facilitate the absorption of excess nutrients in the water and work as natural filters; restoring the "old generation" systems of channels, by avoiding excessive embankments and creating more ways to enhance the natural purification of the river basin.

For example, in Saca di Goro, purification basins on the edge of the industrial plants all along the river should be built.

- After having the debate with a person representing the agriculture lobby, we, as citizens could see that our arguments differ from the lobby ones.

  Indeed, the lobby pretends that the organic conversion will cost the farmer a lot which we can't fully acknowledge: conversion is possible and after having listened to the testimony of a French farmer, we realized that in some cases it won't cost more on the long term.

  We are in the position to create a counter-argumentation to the argumentation of the lobby. We wonder why so much money is invested in GMO, which are not popular amongst citizens, and we suggest that instead, the European Commission and governments focus their *investments on sustainable types of farming* and support to farmers going in that direction.
- To reduce the level of phosphorus and reach better water quality in the Gulf of Riga we need to modernize,the sewage systems and the treatment. In the Goro case, there is also a need of renewing waste water treatment plants, to reduce the level of nitrogen (as a recent national directive pushed for phosphorus removal in the detergents).

II. HOW CAN WE IMPROVE THE CONNECTIVITY BETWEEN RESEARCH, POLICY-MAKING AND CIVIL SOCIETY?

The AWARE project assumes that better connectivity is needed between research, policy-making and civil society to improve marine and coastal water ecosystem management. We fully endorse this position as we observed and experienced important gaps between all stakeholders, gaps that will jeopardize the achievement of sustainable water management in Europe.

Therefore we worked first on the connectivity issue, making recommendations on how to improve the situation through concrete water management actions or EU regulations (1). Our discussions then led us to believe more communication and awareness raising actions were needed towards key targets among stakeholders (2).

1) Connecting actors

Through our experience in the AWARE project and the numerous discussions we had, we found out that we could explain the lack of connectivity between all the concerned actors of coastal water ecosystem management according to two categories of gaps:

- a gap at the level of people
- a gap at the level of countries.

a) Gap of communication between people

Our findings

Our experience is that there are connection gaps between different stakeholders (decision and policy-makers, scientists, business representatives, farmers, fishermen and citizens). Decision makers are often far from citizens and scientists are working on their research; actions are divided or very direct benefit oriented. When the industries are interested in improving their products, they find, obtain, and rely on the scientific information they need. Our experience is that in many cases, when science or politics would like to approach businesses (like shipping companies) on environmental issues, then connections are missing, because the latter find environmental measures/regulations/suggestions expensive and will not implement them spontaneously (with no incentives). In the end this would harm the environment if connections are missing between science and business and the latter does not know about water quality issues.

Although politicians, business and scientists manage to communicate with each other through existing channels (like in the Science-policy interface) we found that *citizens are often left out or information about water quality issues does not reach them*, possibly because information is available in a format we can't understand. Indeed, people are not interested in technical presentations or large quantities of information. To improve communication and to increase connectivity with people, the facts should be presented in a clearer and simpler manner, so information would be more accessible and understandable by those who are not experts in the field.

There is still a gap between technical scientific knowledge and what we know about the actual situation on-site. We find that *feedback from the research and from practitioners should be used more to improve connectivity options.* Indeed, a fisherman told us for instance that he feels better informed when a scientist is on board his ship. He wishes he was consulted as a marine professional. What he observed happened to be at times in contradiction with scientific findings. Overall, this fisherman feels more responsive to the issue of the marine water quality deterioration when one explains to him that these waters are polluted by the watersheds, mostly independently of his actions. He attests to the fact that while being included in a scientific process, he is more attentive to the whole pollution chain and to the role of everyone, his own as well as other farmers', and feels more inclined to act together with others.

Our recommendations

- ***Ambassadors*** – more often high-level officials and popular characters/persons should be used as interpreters to introduce important water quality topics to other lobby groups, business unions, politicians and people.
- ***Local events and festivals*** – can be used to inform citizens, as these are popular gathering places and target people. This is a real possibility to invite local authorities and their representatives, scientists and other stakeholders, who can present their positions or describe their activities. Such opportunities create a chance to have a look into other personś point-of-views.

b) Gap of regulations between countries

Our findings

We have learned that there are different environmental criteria for common geographical regions. For example in the Gulf of Riga there are different regulations about transitional water and coastal water in Latvia and in Estonia, which has an impact in the definition of good ecological status of waters in the area. This causes different activities on the same water basin, which could be harmful for the nature. Also in the Scheldt-Somme-Seine river basin case there were examples of connectivity problems, which were caused by differences in legislation.

Our recommendations

Therefore we find that there should be more *trans-boundary approach* taken into account in the Water Framework Directive. The River Basin Management Plans has just been adopted so we cannot see yet if it will work, but such a trans-boundary approach would definitely better improve connectivity between states and take into consideration the whole water basin. Europe is a common union and therefore trans-border activities should be encouraged. In this case water quality issues offer such a chance.

It would also make sense *to push for such an approach with non-EU member states* concerned by a water basin, like around the Gulf of Riga. *What can the WFD do for that?*

2) Spreading information

We feel the need to inform people on the phenomenon of eutrophication at the sea but also in the countryside, because we are now aware that they are all concerned by the quality of coastal water. It is not only at the sea that the water is deteriorated, but also because of what happens more upstream in the water basins.

Given that, we have decided to define some targets to enhance awareness:

a) Scientists

There is information, but it is not always easy to understand it and not always easy to know where to find it: there is much information we are not aware of. The scientific work and results are very often too complex and "hard to digest" for citizens. Therefore, we recommend the scientists to help and prepare information for different public:

- We ask them to make information for education: that is create folders for children.
- We ask them to be clear in communication to the policy makers to inform them about a problem like the one of eutrophication and so that policy makers can take enlightened decisions.
- Scientific experience and consultation should be used in any case when necessary.
  Often scientists obtain knowledge that can reduce impact on environment, and in cooperation, a safer way of development can be achieved.
- We ask scientists to enter into a dialogue with the citizens, this means inform them, listen to them as experts and give them feedbacks.

Why not doing it in a comprehensible/simplified language? Indeed, a problem like eutrophication needs to be comprehensible for everyone. On our first meeting in Paris we had many presentations by scientists who showed us numbers and diagrams, but we could not understand what all those facts and figures meant. Only after a long questions and answers session we could really understand what they wanted to tell us. Unfortunately this kind information, overloaded by data, is hard to understand and so people lose their interest in it and do not even want to ask, as they are afraid they will not understand.

Also, thanks to what we experienced in AWARE, we have the impression that having discussions and research-collaborations with citizens, permit to shorten the time needed for "round-trips" between research and civil society, and contribute to mutual enrichment and understanding.

Besides, why not letting citizens themselves provide new inputs or directions for further research?

- Expert's opinion should be also up-to-date and flexible so it can be adjusted accordingly to the change of situation.
- We feel that there is a necessity to have some kind of scientific ambassadors to serve as point of contact and "translators" between scientists and citizens or business representatives or politicians alike.

b) Decision and policy makers

We feel they are not enough aware of problems like eutrophication. That's why we ask them to listen to citizens and scientists before creating policies that have to do with the quality of water. They are the ones who can give an impulse to change (i.e. support in general, incentive, education, control). We ask them to be transparent and give us the information of what they are doing and the real results of it. They could also help us with centralizing data and all kind of information about the water quality.

c) Farmers

We would like to convince them to think about changing their way of farming, because the current way (intensive way of farming) has a bad impact on water quality in general, coastal waters quality in particular. We would like therefore to reach the *farmers organizations and lobbyists* so that they can influence their members and bring them to change their way of farming. We call for farmers' organizations to take into account that organic farming is not always expensive and already exists with success.

We have the impression that there is a gap between what farmers' organizations think about organic farming and what we experienced by discussing with farmers who did convert totally or partly their exploitation to organic farming, and got benefits out of it. We understand that all lands are not equal for the conversion to organic farming – as already said - but at least, the issue should be correctly raised and discussed in the farming sector, out of any influence from the agro-industry. We also saw that without necessarily pushing to the conversion to organic farming, actions which look simple can be taken, like producing other kind of plants to make the soil living and filtering.

That's why we recommend targeting farmers and their representative organization at European and local level in doing awareness campaigns.

d) Fishermen

We invite them to share what they experience at sea, because their knowledge is very important. This means sharing with scientists and with farmers, they could learn much from each other. We would also like to make them understand that the way they act as fishermen has an effect on the quality of seawater in general.

e) Children

We'd like to inform them about what is going on, because we want them to act and make them responsible *now and for the future*. We would like to do this by asking the authorities responsible of education in all countries to rethink about education themes. Example: how can we prevent pollution of water; teach children what food to buy; show them how farming can be respectful of the environment; learn them to respect more the environment…

f) Tourists organizations (= policy makers and private initiatives)

We ask them to inform the tourists that there is a problem with the quality of the coastal water, not to scare them, but to really consider them as a partner of the quality of coastal waters: to make tourists conscious of their impact and thus, their role as citizen and consumer when they come back home.

CONCLUSION

We decided to take part in this project because we feel that citizens are often too far away from the discussions impacting their daily life, that is definition of the agricultural practices, help and support of the socioeconomic development of some regions, approval or reject of certain research programs.

We are the ones, by our vote, who give a mandate to policy-makers who are keeping us too far from the policy-making process. We are also consumers who make decisions every day based on the knowledge we have, on buying goods or services of any kind (food, drugs, clothes, mobility means).

One sometimes says "ignorance is bliss". But we want to know… We realized that, when receiving information, we are able to tell precisely what we are ready to do, what we are not ready to do, and explain why it is so. Before that, because we did not know anything scientific about the issue, we could not tell anything about our potential engagement to solve it. However, we were concerned by the issue. Giving accurate information to citizens about an issue allows shedding light on it, allowing them to engage themselves to bringing a solution. *Keeping us, citizens, in the dark prevents us from making full use of our ability to contribute to the decision-making process.*

We are ready and willing to act, to act NOW. Good intentions will not create a sufficient change to win the ecological challenges our world faces, but well thought actions that are initiated now at all levels – especially the European and local ones - supported by strong leadership and motivated actors will bring this change. We as citizens are willing to take the first steps and hope other parties will follow.

ACKNOWLEDGMENTS

We strongly would like to thank all speakers who discussed with us and permit us enrich our opinions and recommendations, as well as all people who contributed to the AWARE initiative:

- ***Local Workshop and Conference of the Gulf of Riga Case Study:***
  Rolands Bebris, Director of the Environmental Protection Department, Ministry of the Environment, Latvia
  Sandra Biedre, Environmental Investment Fund, Latvia
  Andreas Bryhn, Uppsala University, Sweden
  Peter Dimberg, Uppsala University, Sweden
  Maira Dzelzkalēja, Farmers Union, Latvia
  Laura Jankovska, Ministry of the Environment, Latvia
  Atis Minde, Food safety, Animal Health and Environmental Science Institute, "BIOR", Latvia
  Bärbel Müller-Karulis, Institute of Aquatic Ecology, Latvia
  Rene Reisner, the Ministry of the Environment, Estonia
  Inese Šīrava, Tourism Development State Agency, Latvia
  Per Stålnacke, Bioforsk – Institute for Agricultural and Environmental Research, Norway
  Valters Toropovs, Baltic Environmental Forum, Latvia
  Kristina Veidemane, Baltic Environmental Forum, Latvia
  Baiba Zasa, Senior Official, Water Resources Unit, Ministry of the Environment of Latvia
  Kaspars Žūriņš, Rural Advisory and Training Centre, Latvia
- ***Local Workshop and Conference of the Sacca di Goro Case Study:***
  Francesca Andreetti, Institute Delta
  Giampaolo Balboni, WWF Ferrara section
  Cristina Barbieri, Institute Delta
  Davide Bellotti, Councillor of the Ferrara Province
  Silvano Bencivelli, Province of Ferrara
  Andrea Calori, Polytechnic of Milan
  Gianmarco Giordani, University of Parma – UNIPR
  Camino Liquete, JRC
  Alessandro Luè, Polytechnic of Milan – POLIEDRA
  Stefano Martini, Legambiente "Delta Po"
  Chiara Mocenni, University of Siena – UNISI
  Valentino Mondo, Nautica Mondo
  Davide Nardini, Councillor of the Ferrara Province
  Francesco Paesanti, Mussels farmers Consortium of Goro and Portogaribali – Counsellor, Po Delta Regional Park
  Lucilla Previati, Po Delta Regional Park, Emilia Romagna section
  Carlo Sessa, Project Coordinator – ISIS Rome
  Vincenzino Soncini, Mayor of Goro
  Emiliano Sparacino University of Siena – UNISI
  Francesco Tornatore, Po River Basin Authority
  Pietro Valentini, Reclamation Land Authority Ferrara
  Pierluigi Viaroli, University of Parma
  Marcella Zappaterra, President of Ferrara Province
- ***Local Workshop and Conference of the Southern North Sea Case Study:***
  David Alcaud, professor and researcher in political science studies, vice president of ICCR (Interdisciplinary Centre for Comparative Research in the Social Sciences)
  Anne-Marie Banescu, Free University of Brussels
  Gilles Billen, Pierre and Marie Curie University
  Louardi Boughedada, Vice-president of the Urban Community of Dunkerque, in charge of Energy issues related to the Sustainable Development plan and the Climate Plan; president of the Water Local Commission of the Aa Delta

Michel Delebarre, Former State Minister, Deputy and Mayor of Dunkerque City, President of the Urban Community
Nils Fauchon, Veolia Eau
Tomas Garcia Azcarate, expert on Common Agricultural Policy, Institute of European Studies
Josette Garnier, Pierre and Marie Curie University
Erik Gobard, French farmer
Martial Grandmougin, Water Agency Artois-Picardie
Bertrand Hamaide, environmental economist
Francis Kerckof, Beach Working Group (NGOs)
Michael Kyramarios, Federal Public Service Health – Food Chain Safety and Environment – Marine Environment Service
Christiane Lancelot, Free University of Brussels
Anne Lecoeuche, coordinator of the Water Local Commission of the Aa Delta
Alain Lefèbvre, IFREMER Center Channel-North Sea
Marc Lipinski, regional councillor of the Ile-de-France Region
Pascal Maret, Water Agency Seine Normandie
Michel Mariette, Assembly for the protection of the coastal environment of Flandre-Artois (Association Adelfa)
Jean-Claude Martin, director of the WWTP of North-Brussels
Gérard Montassine, retired fisherman
Paul Passy, Pierre and Marie Curie University
Antoine Pierrot, Lyonnaise des Eaux
Bruno Rakedjian, French Ministry of Ecology, Sustainable Development, Transports and Housing
Catherine Ramelot, NITRAWAL-AQUAWAL, association from the Walloon Region)
Pauline Riousset, Pierre and Marie Curie University
Véronique Rousseau, Free University of Brussels
Olivier Rousselot, SIAAP-Public Service of sanitation of the Paris agglomeration
Marc Suméra, director of the Tourism Office of Wimereux

- ***European Workshops:***

Wendy Bonne, European Commission Joint Research Center – JRC
Andreas Bryhn, Uppsala University
Irina Comardicea, Adelphi Research
Peter Dimberg, Uppsala University
Josette Geffray, Université Pierre et Marie Curie – UPMC
Gianmarco Giordani, University of Parma – UNIPR
Liana Giorgi, Interdisciplinary Centre for Comparative Research in the Social Sciences
Fanny Gleize, Missions Publiques
Mathieu Jahnich, environmental communication researcher, SIRCOME
Christiane Lancelot, Free University of Brussels
Camino Liquete, European Commission Joint Research Center – JRC
Alessandro Luè, Polytechnic of Milan, Poliedra
Yves Mathieu, Missions Publiques
Stephen Midgley, International Water Office
Chiara Mocenni, University of Siena – UNISI
Claudia Neubauer, director of the French Citizen Sciences Foundation
Gilles Neveu, International Water Office
Paul Passy, Pierre and Marie Curie University
Pauline Riousset, Pierre and Marie Curie University
Véronique Rousseau, Free University of Brussels
Tania Runge, Copa-Cogeca
Carlo Sessa, Project Coordinator – ISIS Rome
Emiliano Sparacino, University of Siena – UNISI
Per Stalnacke, Bioforsk – Institute for Agricultural and Environmental Research
Nagothu Udaya Sekhar, Bioforsk
Pierluigi Viaroli, University of Parma – UNIPR
Kristina Veidemane, Baltic Environmental Forum, Latvia

IWA Publishing's authorised EU representative for General Product Safety Regulations is Diane D'Arras, 15 rue Duret, 75116 Paris, France, e-mail: safety@iwap.co.uk.

Printed and bound by CPI Group (UK) Ltd, Croydon, CR0 4YY
16/06/2026
02140396-0001